Shaping the Sonoma-Mendocino Coast

Shaping the Sonoma-Mendocino Coast

Exploring the Coastal Geology of Northern California

Thomas E. Cochrane
Geologist

River Beach Press
The Sea Ranch, CA

RIVER BEACH PRESS
Shaping the Sonoma-Mendocino Coast—Exploring the Coastal Geology of Northern California
Thomas E. Cochrane

Copyright © 2017 by Thomas E. Cochrane
All Rights Reserved

Book editor, cover/interior design, and promotion: S.A. Jernigan, Renaissance Consultations (www.MarketingAndPR.com)

All rights reserved. This book was self-published by Thomas E. Cochrane under River Beach Press. No part of this book may be reproduced in any form or by any electronic or mechanical means, including information storage and retrieval systems, without written permission, except in the case of a reviewer, who may quote brief passages embodied in critical articles or in a review.

If you would like to do any of the above or purchase individual or bulk copies, please contact:

Cochrane Enterprises, LLC
PO Box 358
The Sea Ranch, CA 95497
www.RiverBeachPress.com

Published in the United States by River Beach Press
Printed in the United States
Library of Congress Catalog Number: 2016963477
ISBN: 978-0-9985106-0-6
1.) NATURE//Coastal Regions & Shorelines
2.) NATURE /Regional
3.) SCIENCE/Earth Sciences/Geology
First edition

Unless otherwise noted herein, all photographs are the property of Thomas E. Cochrane.

The information provided within this book is for general informational purposes only. While we have attempted to provide up-to-date and correct information, there are no representations or warranties, express or implied, about the completeness, accuracy, reliability, suitability or availability with respect to the information, products, services, or related graphics contained in this book for any purpose. No liability is assumed for damages that may result from the use of information contained within.

Dedicated to all the Sea Ranchers and coastal visitors
who have queried me through the years
about the local geology, including our coast.

A special thanks to my love, Susan Clark,
for her support, numerous suggestions,
aid in editing, and clarification of facts and ideas.

This book is also dedicated to the memory of Janann Strand
who guided me on early trail walks
during which she shared with me her love and
understanding of the natural landscape...

Table of Contents

Preface ii
Acknowledgements iii
Chapter 1 **Getting started on our quest** 1
    Looking at the landscape and kicking over the rocks 5
    Movements beneath our feet—can you SEE or FEEL them? 8
Chapter 2 **Geologic time clock** 13
    Geologic Time Scale: Sonoma & Mendocino coastal area 18
Chapter 3 **Sonoma & Mendocino rock formations** 21
Chapter 4 **San Andreas Fault (SAF)** 31
    Bodega Bay 35
    Fort Ross Area 36
    Plantation 39
    Stream offsets from south of Fort Ross to Plantation 41
    Other structural offsets 42
    Mapping the Black Point Spilite 44
    Point Arena 49
Chapter 5 **Pleistocene Glaciation: cause & development of coastal marine terraces** 53
    Comparison rates of uplift and elevations of California Coastal Terraces 56
    Coastal terraces along Sonoma & Mendocino coast 58
    Terraces on The Sea Ranch 60
Chapter 6 **Special features of interest** 67
    Underground streams 69
    Bowling Ball Beach 71
    Devil's Punchbowl—related sea caves and holes near the Point Arena Lighthouse 74
    (Mendocino County, CA)
Chapter 7 **Coastal river watersheds: Eel, Navarro, Garcia, Gualala & Russian Rivers** 79
    Gualala River watershed resources 83
Chapter 8 **Offshore basins** 89
    Submarine canyons 91
Chapter 9 **Human impacts on our environment** 95
    End notes 97

Appendix: Geologic ROAD LOG 100

At highway stops and beach access points, experience the geology of the coast from Bodega Bay in Sonoma County to Elk in Mendocino County with brief descriptions of unique features at locations along this 85-mile journey

Selected references 135
Index 139
Meet the author 143

Preface

Many things come from the joy of discovery!

For 40 years, since I first discovered The Sea Ranch and wonderful Sonoma-Mendocino coast, I have been looking at this rugged landscape and marveling at the variety of geologic features and processes which are shaping the coast. As I have poked along its inlets and meandering shoreline with maps, rock hammer, hand lens, and Brunton Compass, people often ask me what I'm doing—and a geologic conversation commences.

My core belief is the more we understand the earth beneath us and the natural ongoing processes at work, the more we can enjoy and protect what we have here on our unique stretch of Northern California's coast—and I think we all have a basic urge to understand and make sense of the natural world around us. When geologists look at the landscape, we see more than a rock, a hill, a bluff, a cove, or a beach. In our minds, we see the scope of the entire process which produced that feature, as well as what might happen to it through time. The human mind is the only eye that can see through time and space.

This book began as an update and expansion of Ted Konigsmark's 1994 book, "Geologic Trips, Sea Ranch" which is now out of print but available online on The Sea Ranch Association's website (www.tsra.org). The Konigsmarks left the area a few years ago and Ted graciously gave me much of his library of materials covering this coastal area.

I hope this book will provide you with a reference to the geologic processes which have shaped this region along with a better appreciation of the natural landscape hereabout. The Appendix section, Geologic Road Log (page 100), discusses natural features as well as points of scenic or historical interest where you can see the many specific features which are discussed more generally in the body of the book. The highway stops and Public Beach Access points are marked with highway mile markers as you travel the 85 miles north along California Highway 1 from Bodega Bay in Sonoma County to the small village of Elk in Mendocino County. *Happy trails!*

Thomas E. Cochrane
CA Professional Geologist, License #6124
The Sea Ranch, California, CA

Acknowledgements

I would like to thank the following people for encouraging me to write and finish this book: Barry Richman who helped me in the initial editing process; Ida Egli who read, edited, corrected, and suggested many additions and changes; and my love, Susan Clark, who kept me going in finishing the project. Most of all, my thanks to Sam Jernigan (Renaissance Consultations), who put this all together for me and made this book happen. As this is my first book, I am sure she deserves much credit for her patience in dealing with this new author.

Getting started on our quest

Chapter 1

We have learned so much, but there is yet more to explore—if we look deeply into the depths of nature, what will we discover about ourselves?

The recent explosion of scientific knowledge leads us to an ever-expanding view of the universe, the solar system, our Earth, and the ground beneath our feet. In geology class, we learned, "The present is the key to the past," and erosion and deposition are slow, natural processes. There are, however, large events which occasionally occur—such as a volcanic eruption, earthquake, tsunami, or landslide. Individual humans in the past had only rare contact with these type of events and had little or no knowledge of events which happened on the other side of the world. The telegraph changed all that so the eruption of Krakatoa in 1883 was telegraphed "around the world." The 1906 San Francisco Earthquake was the first big earthquake event to be seen instantly around the world on newly-developed seismographs.

The study of geology developed as a separate science in the early 1800s after Darwin's "Origin of Species" was published in 1859. Mining engineers mapped coal and mineral deposits and named rock formations containing a specific deposit or mineral. Scientists recognized and classified different kinds of rocks, such as sedimentary, metamorphic, and igneous. Rivers and oceans were studied; and in rock strata, geologists identified deposits similar to those contained in old river deposits, beaches, marshes, deltas, offshore bars, and other sites of deposition. Fossils were collector's items, and it was noticed that simpler life forms were embedded in older rocks.

During World War II, technologies were developed on many fronts that helped us better understand the Earth. Oceans were mapped as part of submarine defenses. Undersea mountains, submarine canyons, faults, and various types of sea floor sediments were sampled and mapped. From this knowledge came the concept of plate tectonics which revolutionized the way geologists look at the continents and the ocean basins. The next big jump in knowledge came from the space race. Computers and calculators shrank in size, in cost, and became available to everyone. Satellites mapped the weather and the oceans, and gave us pictures and maps of everyplace on Earth. Now we are mapping the planets in our solar system, and the universe is beyond.

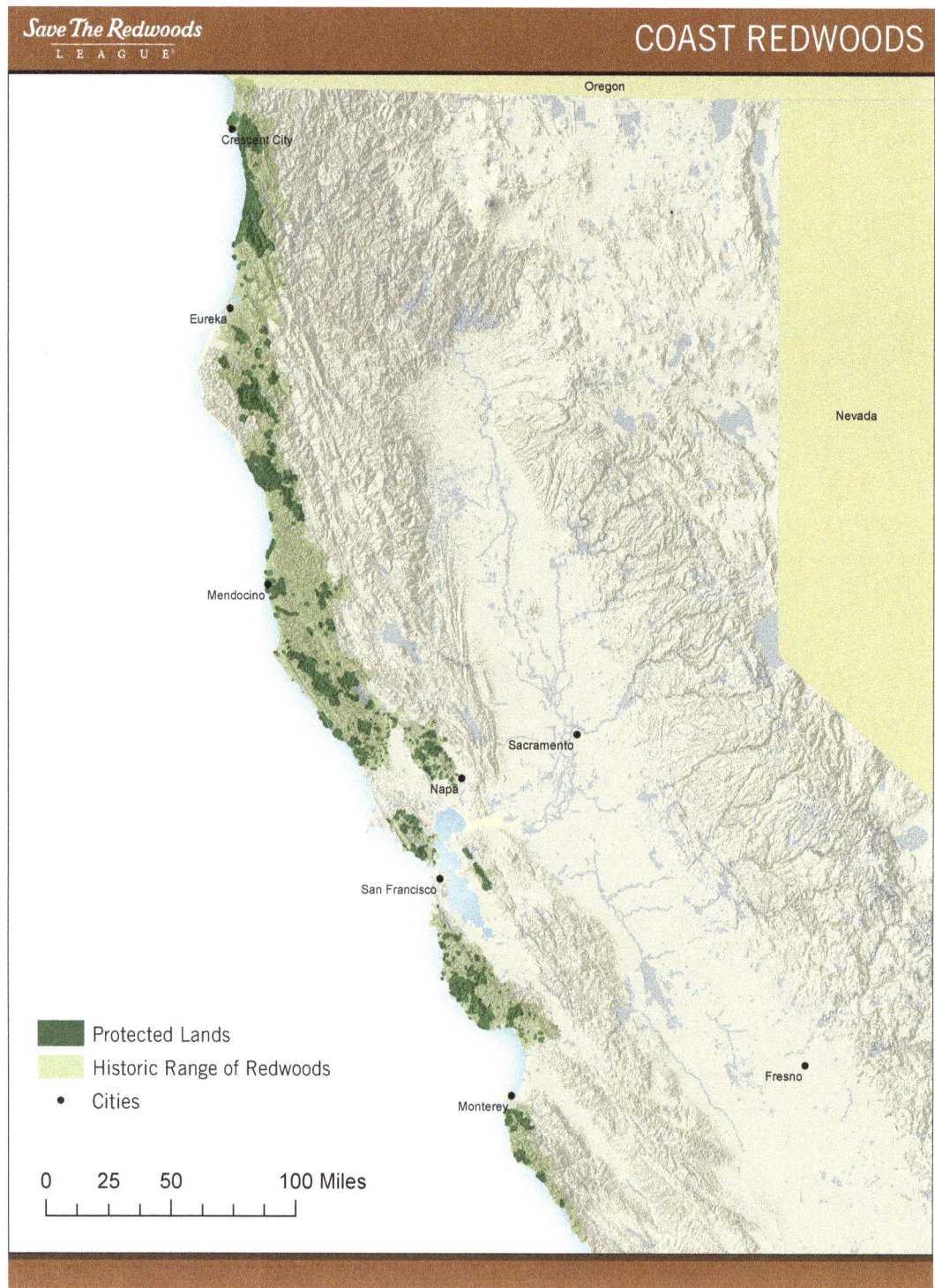

Reprinted with permission from Save The Redwoods League

While in this book the focus is on today's geographical and geological wonders along this 85-mile stretch of the Sonoma and Mendocino coastal area, we will also be keeping in mind distant times and faraway places which have also played a role in the shaping of this much-loved coastline of ours. The coast has been impacted by past climatic events (glacial and interglacial times), by the movement of the North American Plate over the Pacific Plate, by the movements along the San Andreas Fault, and by human activities.

Whether we are visitors to the Sonoma-Mendocino Coast or residents, where should we begin our quest to better understand and appreciate the geologic history of this unique region? Mind you, this is not a boring stretch of coastline with layer cake rocks, slow geologic processes of erosion and deposition, placid waters and gentle breezes as it might first appear.

Thinking back to the mid-1970s when my family and I first discovered The Sea Ranch's 10 miles of coastline, what first attracted us were the magnificent redwood trees. I spent many hours walking the forest trails for the simple wonder of the experience. Then the questions crept in. Where else do these mighty trees grow, and why here? What is the range of the redwood forest? Why do they only grow in the western part of the Gualala River Basin? How large is that watershed? Then we became aware of logging, logging practices, human effects on the landscape, and the declining fishery. And we suddenly became environmentalists!

After a winter storm.

Escaping here to relax and recoup from the demands of our professional lives, the ocean was a big draw, possibly an equal draw with the redwoods themselves. We sat for many hours in the sun with a good book which barely got read, just looking out at the ocean. Ann, my late wife who came from the plains of Kansas and had only recently discovered there were oceans, said the waves reminded her of the waves of wheat rippling in the wind back home.

For days/weeks/months, we spent our time exploring the bluff trails. We saw the migrating whales, seals and sea lions, sea birds, tide pools, and richly diverse sea life. Someone took us for abalone picking at low tide. The ocean had us hooked! Consequently, we bought all the nature books we could find on the coast and began the quest to more fully understand what we were observing and so enjoying.

Our next visit to the coast was in the winter. What a difference—what happened to the gentle ocean? We had never seen such dynamic winds as these, day in and day out! Living in the Midwest, we were aware of tornadoes, but few people experienced them directly. The mighty waves in wintertime were marvelous so we got rain gear and hiked to favorite spots on the bluff edge with the biggest waves in the wildest storms. We reveled in the blustery-but-contained fury of nature!

As we experienced more of the area, new questions arose. What happened to the sand on the beaches? Why does the sand stay on some beaches but others become giant fields of rolling rocks in the winter? How much erosion is happening on the bluffs? Should we build a house on a beach lot? How fast is the bluff eroding?

One day, we felt our first earthquake tremor! It wasn't much of one, but we didn't have them in the Midwest or on the East Coast where I was raised. Once again, there were new questions. What are earthquakes? What is the San Andreas Fault and where is it? When did it move last and when can we expect new movement? Where can we see it? Are all these cracks along the bluff related to the San Andreas Fault? Are they moving? We got all the books we could find and started studying.

We learned this is a dynamic coast with lots going on. As a result, we could not ignore what was happening—or may happen—from natural events and processes. We also could not ignore the effect of humans on the natural environment here regionally. We became *stewards of the land,* committed to doing our best to preserve this lovely coastal environment. In the 1960s, The Sea Ranch adopted the philosophy "to live lightly on the land," which is actually an impossibility, but we nevertheless dedicated ourselves whole-heartedly to living as lightly as possible.

Since then, my studies have expanded beyond just looking at a small section of the California coast, especially as events elsewhere in the world have had effects on this coastline as well. Climate change is causing a slow rise in the ocean level, a change in ocean currents and temperature, and a measurable shift in weather patterns. Animals and plants in Europe are moving northward at the rate of 20 miles per decade. The ocean off our shoreline is growing warmer, and certain species are already dying off or in decline.

The family and I have always found satisfaction in just observing nature and enjoying a time of rest and rejuvenation when we visit our coast. But in acquiring a deeper understanding of how it all works, we now embrace our role in preserving it. Were we put here to use, dominate, and destroy the Earth? Or, do we instead have an obligation to respect and preserve the natural environment? It's my deeply-held belief the latter is true...

Looking at the landscape and kicking over the rocks

We begin our quest...

Geomorphology is the study of land forms. Our coastal landscape is an amalgam of separate distinctive topographic features. The most obvious feature is the Coastal Range of mountains. These mountain ridges primarily run parallel to the coast. A detailed look at the rock strata tells us why this is so: the rocks are folded and faulted, primarily perpendicular to the coast as a result of the pressures of the North American Plate against the Pacific Plate.

Coastal streams are slowly eroding the rock formations. Many of these streams follow faults which run parallel to the coast. The Gualala River mainstem and the Garcia River mainstem follow the San Andreas Fault zone. At the coast, the streams flow off the most westerly coastal ridge directly into the ocean. These streams, known as *consequent streams*, were formed as water running off the rising coastline.

Tributary streams cut across the coastal ridges to join the major rivers. Topographic maps or an aerial photo will verify the stream pattern is like a trellis. There are also some streams or portions of stream watersheds which don't seem to relate to the faulting or folding of the rocks. The Russian River cuts across the landscape in a meandering pattern. The Gualala River's lower section—from the Green Bridge two miles east of Gualala on Old Stage Road to its mouth—has a meandering pattern across the rock structure. It is thought by geologists that these patterns represent previous courses of the rivers which were downcut into the topography as the land uplifted.

The big draw of the area is the ocean and the shoreline. This Pacific shoreline is much different from some other coastlines, such as those along the East Coast of the United States. The East Coast coastal plain slopes more gently into the sea with long sand beaches, and it's many miles inland to the nearest mountain or ridge. By contrast, along the Sonoma-Mendocino coastal area here, we encounter a bluff edge with little or no beach area. The bluff height ranges from

30' to more than 200' in height. Our beaches are short stretches of sand and primarily found in coves. These coves and rocky points produce our spectacular scenery. Coastal features include sea stacks, sea caves, small waterfalls, sand dunes, rocky beaches, and sand bars blocking the mouths of major streams.

Another feature likely to draw your attention are the coastal terraces and much of the area has one gently-sloping terrace just above the ocean bluff. It varies from a few feet in width to nearly one-half mile in width. The landscape above the first coastal terrace also contains eroded terraces which are older and are difficult to detect as erosion has destroyed much of the original terrace in places. The clues to their existence are small topographic benches and old deposits of sand and gravel which were previously deposited on beaches. For long stretches of Highway 1 north of Bodega Bay through The Sea Ranch, the highway is located on Terrace II, overlooking the broader and better-defined lower Terrace I. At many highway stops, you can find beach sand and gravel on the surface, marking a previous sea level high stand. (See Chapter 5.)

Rock types

As visitors kick over the rocks in the streams, along the beaches, and on the landscape, they will note there are many different kinds of rocks hereabout: different types of rocks and different age rocks. Geologists give rock strata names so that when we talk about a given formation by name, we also convey a lot of previously-discovered information on that type of rock.

Along much of our coastline from Fort Ross to Alder Creek, the most abundant rock types are *sedimentary rocks* consisting of sandstone, shale, and conglomerate. These rocks were deposited from older eroded rocks by streams, waves, and turbidity currents in a deep marine basin. Some of these sediments are poorly-sorted muds and silts which have been lithified into mudstones and siltstones. Windblown sands have rounded, frosted grains of very similar size, and these are found on coastal beaches and adjacent bluffs. We don't find any rocks in our area that were windblown, like the cross-bedded sandstones found in the Grand Canyon. However, several small dune areas have been formed along the beach areas near Bodega Bay, the north end of The Sea Ranch, and north of the Point Arena Lighthouse on Manchester Beach.

The *conglomerates* consist of pebbles, cobbles, and boulders (which are terms relating to a rock's size). Conglomerates were deposited in old streams with a steep gradient, or by waves on rocky beaches, or in deep ocean from turbidity currents. The individual pebbles or cobbles may be composed of any type of rock. The Gualala Formation has cobbles of granite and quartzite thought to have been derived from the southern Sierras and deposited in an offshore basin. The movement of the Pacific Plate along the San Andreas Fault has brought the formation 225 miles to as much as 350 miles north from the original site of deposition.

Metamorphic rocks are pre-existing rocks that have been buried to great depth, possibly subducted under the North American Plate, and partially melted or recrystallized into a metamorphic rock. Limestone is metamorphosed into marble, sandstone becomes a quartzite, shale becomes slate and gneiss. Granite rocks are completely recrystallized mixtures of pre-existing rocks. There is granite at Bodega Head. Traveling north along the Jenner Grade, one will notice large dark-colored rocks protruding above the ground surface. These are low grade *greenschist clinkers,* formed by pressure and heat from the squeezing of sandstones and shales along fault zones as the coastal range was faulted and folded. The Black Point Spilite is a slightly metamorphosed igneous basalt.

Igneous rocks were formed deep within our planet by complete melting of pre-existing rocks or from magmas extruded to the surface from the mantle of Earth. In the area along the Jenner Grade and inland along Highway 116, there are thin igneous dikes or sills in the Franciscan Formation. Near Iversen Point, 10 miles north of Gualala and extending north to Schooner Gulch, there's a thin layer of basalt dipping very steeply into the bluff. This rock formation, the Iversen Basalt, was originally a series of lava flows on the ocean bottom. The flows appear to be 15' to 25' in thickness and many of the units have a pillow-like structure attesting to their formation in water.

The Black Point Spilite was an igneous basalt formed under the ocean. A couple of spots along Black Point beach show pillow lava structures attesting to having been deposited in water. The formation was pushed partly under the North American Plate, and partially metamorphosed. The whitish green spots contain chlorite, epidote, and pyrite. The spilite is dark green to black, and weathers to a rusty brown color. The black sands on the beach contain magnetite. The formation has been uplifted as an anticline and is the oldest rock on The Sea Ranch west of the San Andreas Fault.

The surface of the ground is covered with a thin soil that was eroded and weathered from the underlying bedrock formations. These soils vary from clay, to sandy loams, to pure windblown sands on the coastal dunes. The clay soils become very hard and sometimes only support a "*pygmy forest*" when dried-out through the long summer season. The black color found in some soils indicates organic material from old forests and grasslands. Immature soils contain lots of angular, rocky fragments.

Movements beneath our feet… can you SEE or FEEL them?

Wow! All of this motion makes me dizzy! What happened to a flat, quiet Earth?

I receive many questions from local residents about the landscape, erosion, fault movements, plate tectonics, etc. hereabout. This is my attempt to disseminate some complex data in hope of providing you with a better appreciation of our planet and what is happening to the land on which we live.

Earth is currently spinning (rotating) at the dizzying speed of 1,039 miles per hour. It is also racing through space in our annual orbit around the Sun at a much greater speed yet of 66,622 miles per hour. Can you feel it? The universe is racing apart in all directions, although its speed may actually be slowing. When you observe a star or a galaxy in the night sky, its location is not where it appears to be because light only travels at the relatively slow speed of 186,282 miles per second and the closest star is light-years from us plus rapidly moving further away. By the time the light gets here, the star has traveled to a different place. If our sun blows up, we won't know about it for nine minutes!

In the *Deep Time* sense, there is much happening very rapidly (at least as we geologists perceive it) here on the northern coast of California. At The Sea Ranch and Gualala—or, more broadly, from Fort Ross to Alder Creek just north of Manchester—we are sitting on the Pacific Plate, which is moving at the rapid pace of one-half to one inch per year to the northwest along the San Andreas Fault. Actually the surface rocks are locked, and the movement is occurring beneath your feet five or six miles below the surface. Pressure has been building at the surface for the past 110 years since the big 1906 San Francisco Earthquake event when we moved from 8'– 22' in a few seconds in one big shaking event. Since that significant event in the early twentieth century, the plates have moved 6' - 12' at depth, so the question is: when will the surface catch up? Small earthquakes actually occur every day, but we rarely even feel them; whereas the big earthquakes are thought to occur in 100 to 300-year cycles, so we *may* be due—or, not for another 200 years! We surely hope for the latter—*so should I buy or cancel my earthquake insurance?*

Looking further back in Deep Time to the breakup of Pangaea, 250 million years ago Europe and Africa began to split away from North and South America. This split in Earth's crust was caused by rising heat convection currents located deep within the mantle. This brought lava to the surface of the crust at what is known as the Mid-Atlantic Ridge, and it forced the continents apart. This movement of the continents continued from the past to the present, and will continue into the future. Three thousand miles now separates the continents, and the separation continues to widen. The rate of continental movement is approximately three-quarters to one inch per year. As the Atlantic Ocean grows wider, the Pacific Ocean under Asia shrinks. The recent subduction under Japan associated with the 9.0 earthquake and tsunami there was a huge movement of 8'! North of Cape Mendocino, the plate movements switch from the transform movement along the San Andreas Fault to subduction under Oregon and Washington—and we can expect earth movements in that area in the future.

The land here in Northern California is also slowly rising at the rapid rate of .012"per year, or 12" per 1000 years. This rise is due to heat convection currents rising from the Earth's mantle. Before it began its trip to the northwest, the Farallon Plate and part of the Pacific Plate were forced under the North American Plate. The pressures and heat caused the plunging plates to melt, and molten volcanic rocks were extruded up from the mantle onto the land surface. A string of volcanoes extends from Northern California through Oregon and Washington. On the southern end of this trend, in Sonoma, Mendocino, and Lake Counties, small volcanoes and abundant lava flows cover much of the land.

This process began in the Pliocene, nearly five million years ago, and continued until only 10,000 years ago. There are some small volcanic cones near Clear Lake and Mt. Konocti which are only 10,000 years old and undoubtedly greeted our early Indian neighbors. When the lavas extruded into the lakes, volcanic glass was formed. Arrowheads from this local obsidian were traded over a broad region. Heat is still present near the surface of the earth and is expressed at the surface by hot springs and geysers. The Geysers Geothermal Field supplies Northern California with much of our electricity.

Besides riding north and slowly rising, the ocean is trying to advance on the land and erode its surface—in fact, the bluff edge is being eroded at the rapid rate of one to ten inches per year. Due to the melting of the glaciers in Greenland, Antarctica, and in the higher mountain elevations of the world, the sea level is currently rising at the rate of five or six millimeters per year which equals approximately 20' in 1000 years. Zooming in on the timeframe of the current human lifespan, much happens each year geologically from the storm season to the summer season.

In an average year, the waves from storms erode the beaches while later returning the sand back to the beaches. In keeping annual notes of Walk-On Beach and Clarks Cove, I have measured an average vertical change in beach sand levels of seven feet, and occasionally up to 12'. Consider the power of the waves and the erosion of rolling sand and rocks up and down the beach!

On the shortest time scale, the ocean level (tides) change from three to seven feet four times each day due to the gravitational pull of the sun and the moon. Looking at all of these movements, the daily rise and fall of the oceans is the most apparent natural process. Seasonal storms and hurricanes do have a regional effect on people living on the coasts or living on low-lying islands. Earthquakes and tsunamis are more dramatic events, but are usually concentrated in small areas and are spaced many years apart. The long-term effect of rising sea levels in this period of climate change will impact many cities and low-lying areas, causing significant destruction and, eventually, a shift in population centers.

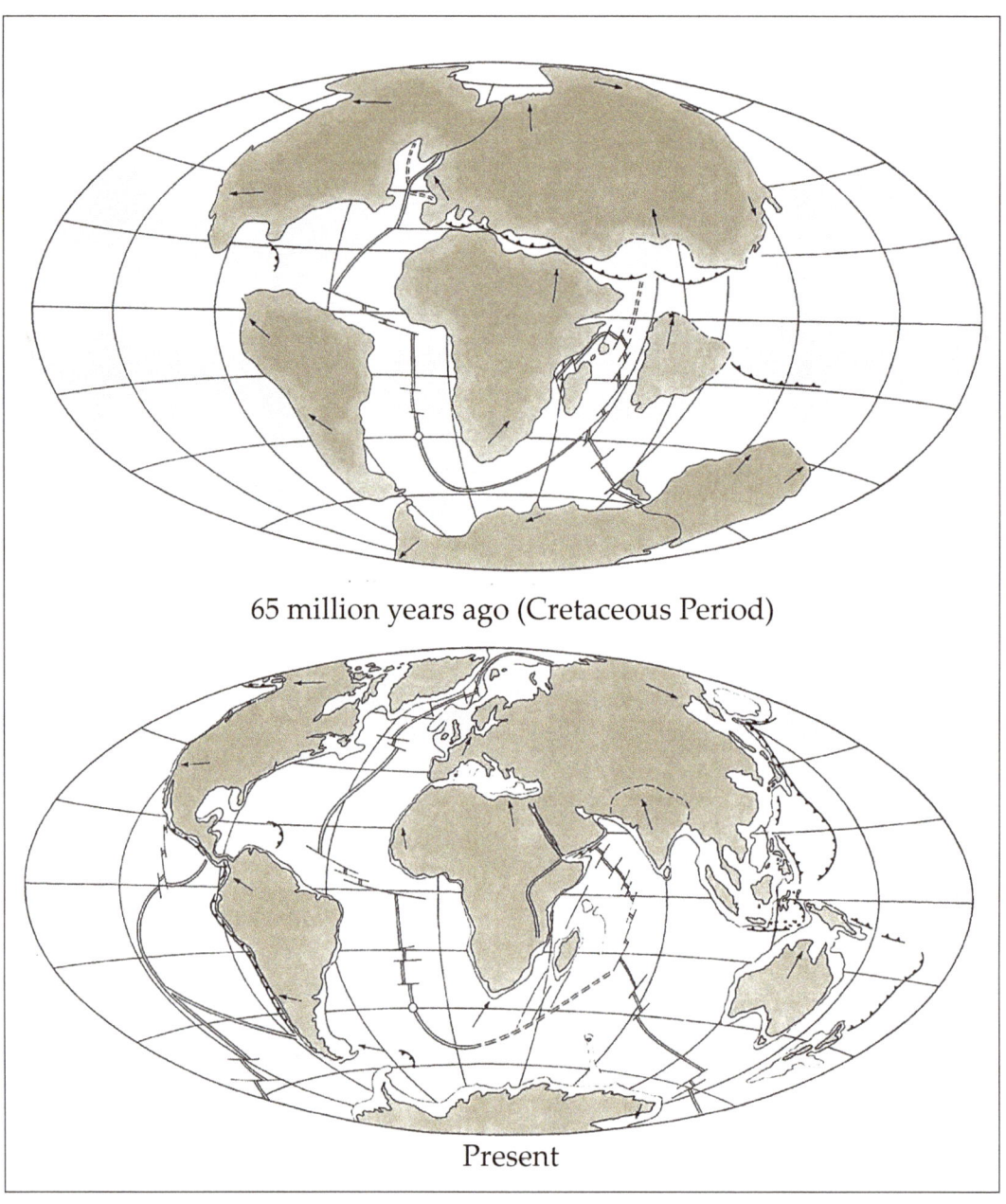

Spreading Pangaea at Cretaceous time and the present configuration. During the Cretaceous Period, offshore basins formed off the California coast from sediments derived from the continent. A meteor impact off the Yucatan Peninsula killed off the dinosaurs and ended the Cretaceous. The San Andreas Fault was not to develop for another 35 million years. Source: Dietz & Holden, 1970

Geologic time clock

Chapter 2

Evolution is part of Earth's history—are humans evolving?

The most difficult concept in studying Earth or the universe is the concept of time. We all measure and conceive of events in human time terms—a generation, the length of time since we were born, when our parents were born, and even when our grandparents era. Beyond these more personal reference points, it is difficult to imagine great lengths of time— **"Deep Time"** (e.g. try to explain to a teenager something that happened more than 10 years ago!).

Over the decades, geologists became acquainted with older and older rocks, fossils, and events. In the late 1700s and early 1800s, geologists began to notice that some rocks contained fossils, and that rocks thought to be older contained fossils with more primitive forms of life. Thus the geologic periods and ages were defined and developed, primarily on the super position of one rock formation succeeding another (older to younger), and based upon the fossil content of those rocks. Modern science methods use Carbon 14 and various radioactive isotopes of minerals to date rocks, and the time scale has been significantly modified and improved from those earliest methods. Events such as the giant meteor crash that ended the Cretaceous Period and killed off the dinosaurs and many other forms of life can now be traced to a thin zone. The specific time attributed to this event gives geologists a marker in the Cretaceous rocks that is valid for worldwide correlation of Cretaceous rocks.

Scientists have now pushed the beginning of the universe, the "*Big Bang*," back to 13.7 billion years ago. Earth is now thought to have formed 4.6 billion years ago. Life forms on our planet have been identified from fossils dating back three billion years. Human-like life forms only evolved around two million years ago. Humans have only been here a very short amount of time compared to the age of the universe. If there is life out there in the rest of the universe, consider that life on Earth began relatively late in time, after the Big Bang, compared to the great age of the rest of the universe. A planet near some other star could have formed several billion years earlier than Earth, and if located in the *goldilocks region* suitable for life, had time to develop life forms which are possibly even more advanced than Earth's humans!

Included on the following pages is a Geologic Time Scale with specific details and facts about life that occurred, or events that happened during those periods. **The Earth is a book, and with the proper tools, we can read it**. Sea water has been trapped in various sedimentary rocks and with different salinities and trace elements, giving us clues as to past temperatures and environmental conditions. Different life forms live in different environments, at different depths, and succumb to different salinities, different pHs, as well as different oxygen and carbon dioxide levels. Coral reefs are rare in the current climate, but in times past covered large sections of Earth, including the polar regions. In earlier ages, during the Carboniferous Period, vast areas of marine marshes or lagoons formed which produced thick beds of coal.

Focusing on our coast, we find that not all of the geologic periods are represented or are found in one continuous sequence. Some rocks were deposited here earlier but have since been eroded. At other times, the continents were situated higher and no rocks were deposited in this area to represent that time period. To complicate the picture, some rocks have been moved around and are now located far from where they were formed. Before plate tectonics, geologists had difficulty explaining why some of these rocks were not where they were expected to be.

The **Geologic Time Scale** represents the measurement of time from Earth's formation 4.6 billion years ago to the present. It's broken down into units, with each succeeding unit representing a shorter period of time. The largest unit is called an *Eon* of which there are four; three of which cover the *Precambrian*, once thought to represent rocks developed before life arose on our planet.

The earliest time is called the *Hadean Eon* and covers the period from Earth's formation 4.6 billion years to 3.8 billion years ago. This is when the atmosphere formed as a result of collisions with asteroids, and our moon was also formed. There are no rocks preserved from this time, if indeed solid rocks were even formed.

Next youngest is the *Archaean Eon* from 3.8 billion years to 2.5 billion years ago. Water began to collect. The atmosphere was 75% nitrogen and 15% carbon dioxide. Solid rocks formed on the surface of Earth, and primitive lifeforms began to develop. Following this was the *Proterozoic Eon* from 2.5 billion years to 542 million years ago. Stable continents were formed. Free oxygen was found in the atmosphere and oceans. Anaerobic organisms died off in the presence of oxygen. Vast precipitates of iron ore formed banded iron ore deposits. The first ice ages occurred, and the variety of simple life forms increased. The *Phanerozoic Eon* was from 542 million years to the present and includes all the familiar **Eras** and **Periods.**

The **Paleozoic Era** (old life) covers the geologic **Periods** ranging from 542 million to 251 million years ago.

> **Cambrian Period** (542 to 488 million years). Multicellular life flourished and the first marine animals with shells appeared.
>
> **Ordovician Period** (488 to 443 million years). Marine invertebrates became common. First green plants appear on the land. Glaciation of the super continent Gondwana occurred with the mass extinction of many marine invertebrates.
>
> **Silurian Period** (443 to 416 million years). Coral reefs appeared. First species of fish with jaws and sharks appeared.
>
> **Devonian Period** (416 to 359 million years). Ferns and seed-bearing plants, first forests formed, and first amphibians appeared. Oxygen level of atmosphere was about 16%. Global cooling created marine extinctions.
>
> **Carboniferous Period** (359 to 299 million years).
>
>> **Mississippian Epoch** (359 to 318 million years). Large primitive trees grew. Oxygen levels increased. Vertebrates appeared on land.
>>
>> **Pennsylvanian Epoch** (318 to 299 million years). First reptiles appeared. Oxygen levels increased to 30%. Huge deposits of coal formed in North America, Europe, and Asia.
>
> **Permian Period** (299 to 251 million years). Formation of super continent Pangaea happened. Conifers appear. Earth was cold and dry. An ice age and large volcanic eruptions deposited voluminous amounts of carbon dioxide, methane, and hydrogen sulfide gases into the atmosphere, causing massive extinctions of 90% of ocean dwellers along with 70% of land plants and animals.

The **Mesozoic Era** (middle life) covers the geologic **Periods** ranging from 251 to 65.5 million years ago.

> **Triassic Period** (251 to 199 million years). Breakup of Pangaea began. Reptiles populated the land, small dinosaurs on land and ichthyosaurs and plesiosaurs populated the seas.

Jurassic Period (199 to 145 million years). Earth warmed. It was the Age of the dinosaurs, and small mammals appeared. The **Franciscan Formation** was formed as the basal rock unit on the Northern California coast, with no older rocks found.

Cretaceous Period (145 to 65.5 million years). Breakup of Pangaea continued. Flowering plants, modern mammals, and birds evolved. Large volcanic eruptions occurred in India. Meteor impact in the Yucatan ended the period with mass extinctions of 80+% of marine species as well as 85% of the land species (end of the dinosaurs).

The **Cenozoic Era** (recent life) covers the time from the end of the Cretaceous to the present.

Paleogene Period (65.5 to 23 million years)

Paleocene Epoch (65.5 to 54.8 million years). Placental mammals, flowering plants widespread, and hoofed mammals appeared. Formation of the Rocky Mountains commence.

Eocene Epoch (54.8 to 33.7 million years). India ran into Asia and formed the Himalaya Mountains. Modern mammals appeared. Global cooling created the Antarctic ice sheet.

Oligocene Epoch (33.7 to 23 million years). Grasses appeared in abundance which slowed the rate of erosion. Elephants appeared.

Neogene Period (23 million years to today)

Miocene Epoch (23 to 5.3 million years). Antarctica separated from Australia and South America. Antarctic ice caps arose. The continental interiors dried. Forests gave way to grasslands. Early hominids (great apes) appeared six million years before the present.

Pliocene Epoch (5.3 to 2.58 million years). North and South America joined and animals crossed the land bridge. Cooler climate, grasslands, and savannas welcomed long-legged grazing animals. Australopithecus (a now-extinct genus of ape) appeared in eastern and north Africa.

Quaternary Period (2.58 million years to today)

> **Pleistocene Epoch** (2.58 million years ago to 11,700 years ago). There were a series of continental glaciations and Interglacial warm periods. Yellowstone had a volcanic eruption. Tool-making hominids evolved two million years ago. Neanderthals migrated to Europe 230,000 years ago.
>
> Cro-Magnon (homo sapiens) man appeared in Europe 40,000 years ago. Neanderthals disappeared 28,000 years ago. Humans arrived in the Americas. The Wisconsinan Glaciation (Younger Dryas in Europe) ended the epoch with sea levels rising more than 300' to the current sea level.
>
> **Holocene Epoch** (11,700 years to present). A warm Interglacial period began and humans populated our planet. The rest, as they say, is history!

GEOLGIC TIME SCALE—Sonoma & Mendocino coastal area

ERA	PERIOD	EPOCH	MYA*	EVENTS & FORMATION NAMES
		HOLOCENE	.011	*Glacial Age ends, sea levels rise*
	QUATERNARY		.081	*Terrace 1 Sea Ranch Meadows*
		PLEISTOCENE		*Sea level 300' lower*
			2.58	*Uplift of coastal ranges*
CENOZOIC		PLIOCENE	5.3	Ohlson Ranch Formation (Por)
		MIOCENE		Monterey Formation (Tmm)
				Point Arena Formation (Tmp)
	TERTIARY			
			23.0	Galloway Formation (Tmg)
			23.8	Iversen Basalt (Ti)
		OLIGOCENE	33.7	
		EOCENE	54.8	German Rancho Formation (Teg)
		PALEOCENE	65.5	
				Meteor impact, dinosaur demise
		UPPER		Gualala Formation (Ka)
				Anchor Bay Member (Kua)
				Stewarts Point Member (Kus)
	CRETACEOUS			
MESOZOIC				Franciscan Formation (Kjf)
		LOWER	145.0	Black Point Spilite (Kbs)
	JURASSIC		199.0	Age of the dinosaurs
	TRIASSIC		251.0	

*MYA = Million Years Ago

This is a photo of layered sandstones and shale found in the bluffs along many of the area's beaches. The rocks at the bottom of the photo are thin-bedded shale and siltstone, indicating slow, deepwater deposition. The lighter-colored rock is a tongue of sandstone formed as a deepwater landslide off the continental shelf which slid into a deep basin. These underwater slides are called turbidites.

The top of the photo is another thicker sandstone turbidite. The rocks above the cell phone are shale and siltstone which were probably also deposited in deep, quiet water. However the turbidite slide churned up the shale and silts into twisted forms. (Hence the name turbidite--e.g. caused by a turbid flow of water.) The vertical cracks are joints in the rocks with no movement indicated. Probably if you look a few feet in either direction, you will encounter a minor fault (sometimes a larger one) in which you will be able to see a few inches or feet of displacement. These are old, inactive faults which formed when the rocks were uplifted or wrenched by plate tectonics.

Sonoma & Mendocino rock formations

Chapter 3

What happened to all the missing rocks and ages?

Here on the western edge of the Northern California coast, the rocks encountered under foot do not represent all of the geologic time periods just discussed. The missing rocks were never here, or they were thrust under the North American Plate by plate tectonic movements, or they were carried north along the San Andreas Fault. Large sections of the geologic column have been eroded during periods of uplift and then redeposited in deep water basins offshore.

For many years (before the concept of plate tectonics), scientists wondered why there were no old rocks in the ocean basins. To my knowledge, none or very few rocks have been discovered in the ocean basins with ages older than the Jurassic Period (199-145 million yrs.). However, on the continents at the far reaches of the globe, rocks from the Archaean Eon which are over three billion years in age have been discovered.

The continents consist of lighter igneous and sedimentary rocks than the ocean basin basaltic crust and therefore float over the denser rocks. Some of the very oldest rocks have been preserved on some portions of the continents, but most have been assimilated by under thrusting beneath the continents. Therefore our knowledge of the older periods of time becomes more interpretive in light of less data being preserved.

Rock Types

There are three basic types of rocks: igneous, metamorphic, and sedimentary.

Most of the rocks in our coastal area are *sedimentary* sandstones, siltstones, conglomerates, mudstones, and shale. These rocks were originally sediments which were eroded from the land by wind, running water, and waves and then deposited in offshore basins. (See the section on the offshore basins, page 89). With time, the original sediments became cemented into hard rocks. Common cements holding the grains together are calcite derived from sea shells and silica derived from sand. Water flowing through the sediments with pressure from the overlying sediments caused the cementation.

Several patches of *igneous* rocks are scattered along our coast. Igneous rocks were formed by extreme heat, often deep inside the earth, and may have melted pre-existing rocks of any type. *Intrusive igneous* rocks were formed deep inside the earth and have come to the surface as a result of erosion of overlying rocks. *Extrusive igneous* rocks have come to the surface of the earth as lava flows or, more violently, as volcanoes. Geologists recognize igneous rocks by looking with hand lenses or microscopes (or X-rays) and identifying crystal faces and different minerals.

Metamorphic rocks are less common in this area. They were formed by heat and pressure that recrystallized the original rocks, but did not melt them completely—as in the formation of igneous rocks. The Black Point Spilite found on the south end of The Sea Ranch is a slightly metamorphosed basalt. From Jenner south to Bodega Bay, large dark grey blue rocks (*blueschists*) protrude from or are mixed in the Franciscan Formation, which is an old sedimentary rock formation. On the east side of the San Andreas Fault zone is an area of thick sediments, the Franciscan Formation of Jurassic/Cretaceous Age. Folding and faulting, which produced our coastal ranges, put tremendous pressure on some areas along the faults. Recrystallization has produced zeolites, blueschists, and eclogites. Serpentine, garnet amphibolites, and glaucophane schists can also be found along Skaggs Spring Road toward Lake Sonoma.

Rock Formations

Geologists group rocks into formations and give them names, which helps them talk to one another about a specific formation. A rock formation usually consists of rocks formed together in a similar environment at a specific time. Some formations may span an entire time period. Occasionally, a formation laps from one time period into another period. The following formations are found in our coastal area.

Black Point Spilite (Kbs) Jurassic or Lower Cretaceous Age

The *Black Point Spilite* is the oldest rock formation found west of the San Andreas Fault on The Sea Ranch. It outcrops along the bluff edge to the north of Black Point to the north end of Pebble Beach, a distance of over 8000'. It is hard, dark green to black in color, and highly faulted and fractured. The beach sand along Black Point Beach is green to black in color, but is also mixed with lighter-colored sand eroded from the overlying adjacent formations.

The Black Point Spilite was deposited as an underwater series of basalt lava flows in a back arc position from an Island Arc series of volcanoes. (The Japanese islands are a volcanic arc series with the back arc position to the west of Japan). The site of these lava flows is thought to have been hundreds of miles to the south of its current location near Santa Barbara, and occurred over 84 million years ago. After deposition, these lava rocks were subducted underneath the North American continent and metamorphosed by heat and pressure to the low grade greenschist facies. Subsequent to this, the rocks have been covered with younger unmetamorphosed sediments. Later, they were uplifted, faulted, and folded into the formation of the California Coastal Ranges. Subsequently, in the last 15 million years, they migrated north into our area via plate movements bounded on the east by the San Andreas Fault.

Franciscan Formation (KJf). Jurassic and Lower Cretaceous

The *Franciscan Formation* may be similar in age to the Black Point Spilite. It is found only to the east of the San Andreas Fault zone. It consists of a series of conglomerates, sandstones, and mudstones deposited in an offshore basin from the rising continent to the east. Subduction of oceanic basaltic rocks under the North American continent caused the rocks to melt, resulting in the Sierra granite batholiths. As the mountains rose, finally terminating in a pulse in the Cretaceous period approximately 80 million years before present (80ma BP), sediments poured into the Franciscan trough, accumulating to as much as 50,000' in thickness. Continuing subduction under the continent caused the Franciscan to be faulted and severely folded. The pressures of the folding and faulting process produced low grade metamorphism along some of the fault lines. The blueschist boulders form prominent knobs along the Jenner Grade and inland in Sonoma County. *Igneous dikes* occasionally protrude from the surface where hot spots occurred, a result of more intense melting or injection of magma from deeper inside the crust.

The Franciscan has been a major contributor to the sediments deposited in the offshore basins along the coast. The only formations of rock preserved that were deposited on the Franciscan in the western part of the coastal ranges are the Ohlson Ranch Formation and the Wilson Grove Formation of Pliocene Age. In the Central Valley, thick deposits of younger sediments cover the Franciscan.

Gualala Formation (Ka) Upper Cretaceous

The *Gualala Formation* is divided into two members, the **Stewarts Point Member (Kus)** and the **Anchor Bay Member (Kua)**. Carl M. Wentworth (1976) measured a minimum thickness of the Gualala Formation at 8000'. The Stewarts Point Member rests as a fault contact on the Black Point Spilite, so we do not know if Lower Cretaceous rocks were laid down on the Black

Point Spilite, or there was a period of non-deposition, or deposition and erosion. The Stewarts Point Member rocks consist of a series of sandstones and conglomerates. The conglomerates have a predominance of granite and quartzite clasts, as opposed to the Anchor Bay Member conglomerates which are darker in color with abundant mafic clasts. The contact between the two member formations occurs on the beach just south of the Knipp Stengel Barn on The Sea Ranch. On the north end of the beach is a resistant layer of conglomerate with dark clasts, and on the south end of the beach is a lighter-colored conglomerate belonging to the Stewarts Point Member. A squeezed shale zone occurs at the stairway, possibly marking the formation boundary.

The contact zone of the Stewarts Point Member and the Black Point Spilite at the south end of Black Point Beach is a 20' fault zone of black breccias (fault gouge) dipping approximately 40° to the south-southeast. The fault strikes N. 60° to 80° E. The steeply dipping Stewarts Point bedrock is a conglomerate with clasts up to 12" in diameter. A thin conglomerate just on top of the Spilite has rounded clasts of Spilite, which possibly indicates the Stewarts Point Member was deposited directly onto the surface of the Black Point Spilite and had not been thrust faulted, thus losing an older section of the formation. The Stewarts point Member extends on the south flank of the Black Point Anticline to just south of Stewarts Point, where it dips beneath a short section of the Anchor Bay Member before both dip beneath the German Rancho Formation.

The Anchor Bay Member contains similar sandstones, shale, and conglomerates, with the previous noted increase in mafic clasts in the conglomerates and more plagioclase feldspar present in the sandstones.

Together, these two members of the Gualala Formation are approximately 5000' in thickness. The original site of deposition occurred possibly 350 miles to the south of their current location. The source of the conglomerates is thought to have come from the Sierras. At the time of deposition, the Gualala Basin (Bodega Basin) was subsiding and many of the deposits are identified as *turbidites,* which were deep-water sediment flows (slides) off the shelf of the continent which were caused by earthquakes, tsunamis, or storms.

German Rancho Formation (Teg) Lower Eocene Age

In his 1966 dissertation (Stanford University), Carl M. Wentworth estimated the German Rancho Formation is 20,000' in thickness. The rocks consist of shale, mudstones, and sandstones with a noted lack of conglomerates. The shale and mudstones are thicker bedded than those in the Gualala Formation, indicating slower, deep water deposition. Lower Eocene fossils have been identified in these rocks. Once again there appears to be a missing section of Paleocene Age rocks. The contact of the German Rancho Formation with the Anchor Bay Member is a fault contact near the south end of Walk-On Beach on The Sea Ranch.

Progressing northward from Walk-On Beach, one encounters sandstones of varying thicknesses and dimensions, which have also been identified as turbidites. Flow features, graded bedding, and erosional fluting of the underlying beds attest to the turbidite flows. The German Rancho Formation extends on the surface to the Gualala River mouth. From Gualala north to Steens Landing, the German Rancho Formation has been eroded on the west side of the coastal ridge between Triplett Gulch and Roseman Creek, exposing the underlying Anchor Bay Member.

The German Rancho Formation, consists of predominantly thin to thick bedded sandstones. It is present at the surface north of Steens Landing and extends north to Schooner Gulch. The Galloway Creek Fault, considered a significant fault which is trending northeast-southwest, is downthrown to the northwest, preserving younger Miocene rocks on the surface.

On the south limb of the Black Point Anticline, the German Rancho Formation is found on the surface from its contact just south of Stewarts Point to Fort Ross. The last mile south of Fort Ross, before the San Andreas Fault moves offshore, the Galloway Formation (Tmg) of Miocene age consists of shale, mudstones, and siltstones. It is situated on top of the German Rancho formation. By stopping at the pullout south of the concrete retaining wall (mile post SON 29.00) and looking north into the ravine adjacent to the wall, one can see Miocene shale in the face of the ravine and Franciscan Jurassic/Cretaceous sandstones at the right side (east), a gap in time of possibly 100 million years and a gap in the space of site of deposition of 350 miles.

Iversen Basalt Formation (Ti)

Jumping back to the north, at Iversen Landing (MEN 9.76) there is a drastic change in lithology. At the shore side are thin bedded sandstones and mudstones of the German Rancho Formation that are uncomfortably overlain by a black basalt. The *Iversen Basalt* consists of a series of lava flows, each possibly 10' or 15' in thickness, and building up to possibly 800' to 900' in total thickness. Some areas show good evidence of pillow lavas. Vesicles (holes that were gas pockets in the molten lava) are common and some vessicles have been filled with white phenocrysts of calcite and feldspar. The Iversen Basalt is considered to be Lower Miocene in age and dated at 23.8 million years BP.

The change in lithology is dramatic at Iversen Landing with yellow and orange sandstone beds on the east, and the black basalt on the west. The rocks dip at a high angle to the southwest. The contact crosses the point at Iversen Landing and runs north, crossing Highway One at milepost MEN 9.76. Public access is available at MEN 10.08 with a trail to the beach and good views of the formation at bluff edge. The formation is terminated at the north end by a cross fault that follows Galloway Creek, or possibly the north fork of Schooner Gulch. The southern end extends beneath the ocean and outcrops as a sea stack called Sail Rock opposite of MEN 8.21.

Skooner Gulch Formation (Tmg(s)

The Skooner Gulch formation is exposed along the sea cliff south of Schooner Gulch. It consists of a series of light brown, clayey sandstones. Early Miocene mollusks have been identified in the basal part of the formation. The rocks dip steeply to the west. Several NE–SW trending adjustment faults cut the formation. Subsequent erosion often exposes the underlying Iversen Basalt along faults in sea caves just south of Schooner Gulch.

Galloway Formation (Tmg(a)

North of Schooner Gulch the Galloway Formation is exposed in the sea cliff. Dips range 40º to 66º to the south-southwest. The rocks consist of thin-bedded sandstones, mudstones, and siliceous shale. The predominant color is a light, whitish grey. These rocks are exceedingly soft and subject to minor landslides. In strong winds, their particles rain down on the beach. Various large concretions up to several feet in diameter are eroding out of the bluff face. The so-called *bowling balls* are aligned along a series of beds at the foot of the bluff. (See section on Bowling Ball Beach, page 71).

Monterey Formation (Tmm)

Charles Weaver (1943) named the rocks around Point Arena and Arena Cove the **Point Arena Sandstone (Tmp)**, which is now considered a member of the Monterey Formation, with its age estimated as mid to late Miocene. The rocks consist of a series of diatomaceous to porcelaneous shale and siltstone, overlain by a series of thin sandstones and sandy mudstones.

Ohlson Ranch Formation (Tor)

Moving inland to the Annapolis area and covering much of the Gualala River watershed, one encounters an orange-brown to orange-yellow sandstone capping the eroded hills. This is the Ohlson Ranch Formation of middle Pliocene Age, deposited around five million years ago. The sandstone reaches around 300' in thickness, and has been eroded by the Gualala River watershed tributaries as the land has slowly risen since the Pliocene.

The sandstones are marine in origin, deposited in a shallow basin similar to San Pablo Bay. The Ohlson Ranch sandstones are found in an area approximately four miles in width and 23 miles in length. In Pliocene time, the connection to the ocean for the embayment was centered around the space now occupied by the area between Buckeye Creek and the Wheatfield Fork. These streams now enter the mainstem of the Gualala River, but were open to the ocean in Pliocene time.

It is important to remember that the San Andreas Fault zone has carried the area on the west side of the fault northward at approximately one inch per year. Thus five million years ago, the Sea Ranch area west of the current Wheatfield Fork junction with the Gualala River was 80 miles south of its current location. Eighty miles to the north on the Gualala Block is currently under the ocean, and 25 or more miles north of Fort Bragg. A small amount of subsidence during Pliocene time allowed the shallow basin to develop. It is thought that the Coastal Ranges inland from the San Andreas Fault have been rising at different rates in different fault blocks. Many of the tributaries of the Gualala River, the Garcia River, and Russian River follow fault traces and folds of the Coastal Range. Most of these faults run parallel to the San Andreas Fault. Some of them also exhibit strike slip movements, and a few are still active. (i.e. Healdsburg Fault).

Since the time of deposition, this whole area has experienced uplift—but at different rates in different blocks. Near the Wheatfield Fork, the contact of the Ohlson Ranch Formation and the underlying Franciscan Formation is at 360' elevation. Near Buckeye Creek, the contact is at 1200' elevation, and at Table Mountain east of Fort Ross, the contact is reported at an elevation of 1750'. Calculating the rise of the land from sea level to 360' elevation in four million years, we get a rise of .001 inches per year; to 1750' elevation, we get a rise of .00525 inches per year.

If we calculate the rise of the land west of the San Andreas Fault, Terrace I was formed 80,000 years ago by wave action, and is now situated at 100' elevation on The Sea Ranch. This calculates a rise of the land at the rate of .015" per year. North of Alder Creek where the San Andreas Fault leaves the land, Terrace I is at 200' elevation, which translates to a rise of .03" per year.

Quaternary sediments

The coastal terraces are partially covered with unconsolidated to poorly cemented sands and gravels deposited during Pleistocene periods of high sea levels. They predominately consist of old beach sands and gravels, with some areas of windblown sand. Stream deposits cover some of the terraces adjacent to major streams. The lowered valleys of the major streams—particularly near their mouths, such as the Russian River, the Gualala River, the Garcia River, and other streams to the north—are filled with up to 200' of sediments deposited during periods of high sea levels. These stream valleys were cut deeper by over 200' when sea levels were lower by as much as 300'.

Black Point Beach, The Sea Ranch, CA

Note the undulating surface of the dark-colored sandstone which dips into the bluff. The wave cut surface was covered with sand and gravel when the ocean retreated. These rocks are 70 million years old. The overlying sand and gravel was deposited 80,000 years ago on the wave cut terrace formed 81,000 years ago. The tan sand is windblown sand currently forming. Current wave erosion is forming sea caves.

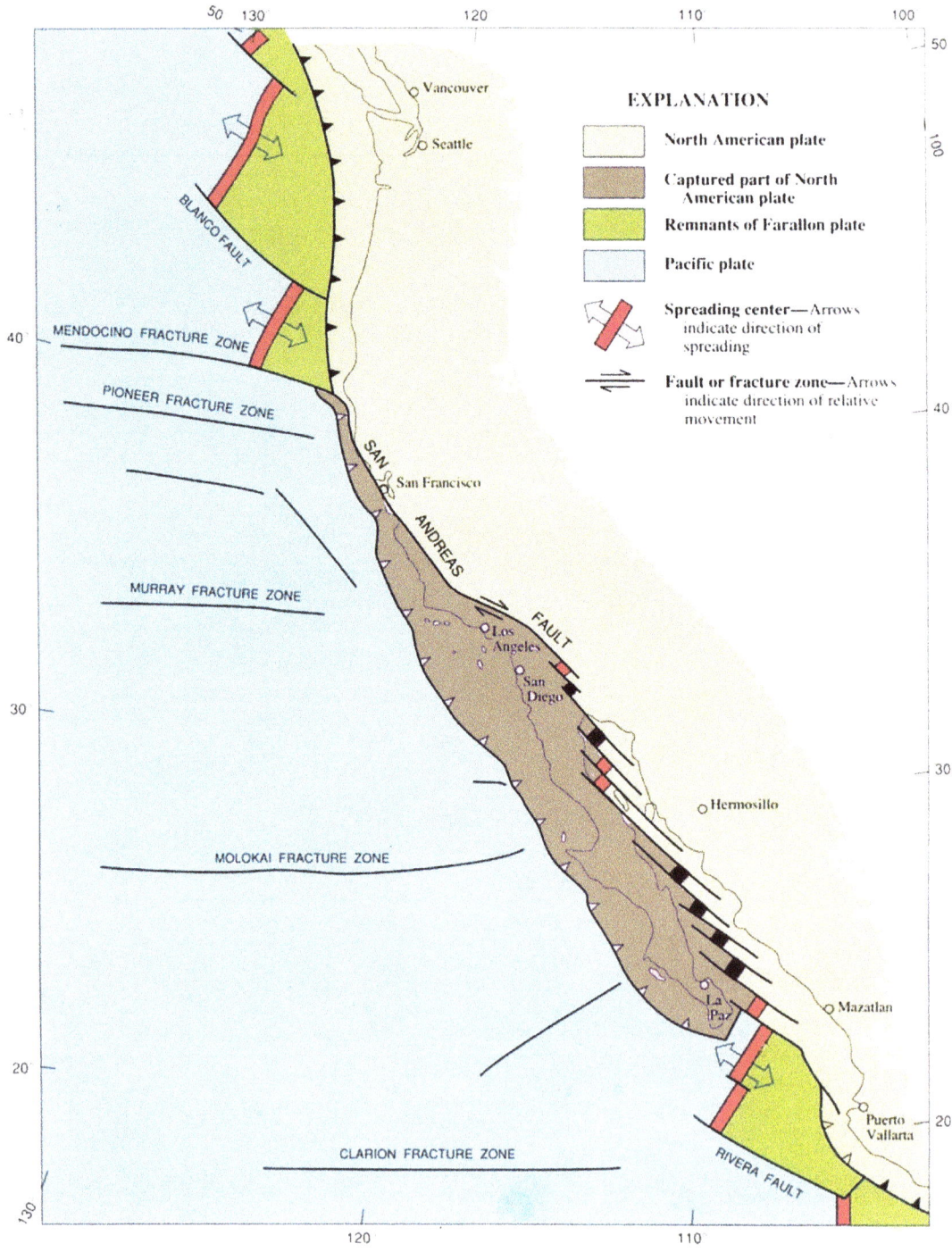

Source: US Geological Survey, 1990, Professional Paper 1515

San Andreas Fault (SAF)

Chapter 4

Is it moving? And when might the Big One hit?

California might be called the land of seismic faults for we seem to encounter them throughout the state. The mountains have been thrust up along these faults. The large basin areas such as the Great Valley, the Salinas Valley, Santa Rosa Plain, Napa Valley, Round Valley, Redwood Valley, Clear Lake, and many others are all bounded by faults. Many streams are eroded along fault traces. Our coastal area offers many opportunities to see the effects of old faulting exhibited in the coastal bluffs, as well as more recent movements along the San Andreas Fault (SAF) as demonstrated by offset geologic formations, offset streams, sag ponds, trees, and other features.

Most of the faults we can see are ancient fault traces, many thousands to millions of years in age, although the effects of these faults still control the shape of the landscape. Some of these faults still experience movement and can produce earthquakes.

The San Andreas Fault became well-known and was thoroughly studied following the 1906 San Francisco Earthquake. For the first time, seismographs around the world had registered the earthquake, and this resulted in subsequent studies which provided insight into earthquake waves, and even gave us new information about the core and mantle of the planet.

The map of the west coast on the opposite page illustrates the broader picture of the movements of the adjoining plates. The North American Plate is overriding the Pacific Plate the entire length of the map. The lavender-colored area, which includes our area at the north end, is really part of the North American Plate, but has been broken off and slides north against the North American Plate. It is bounded by the San Andreas Fault on the east side for 850 miles, from the Mendocino Fracture Zone to the Sea of Cortez next to Baja California. Note the spreading zone faults also have displacement to the north. The Farallon Plate is still being subducted under the North American Plate offshore of Washington and Oregon.

The San Andreas Fault Zone is the boundary of the North American Plate and the Pacific Plate. The Pacific Plate underlies the Pacific Ocean and contains mostly dense balsaltic rocks in a thin crust. The North American continent is comprised of lighter rocks and these ride over the denser oceanic crust. The small piece of land that is west of the fault in our area is referred to as the Gualala Block or the Salinian Block. These continental rocks are deposited in deep subsiding offshore basins. If the San Andrea Fault had not formed—dragging us to the north—the Gualala and Salinian Blocks would have been subducted under the continent, melted and extruded as volcanoes like those in Washington and Oregon.

The south end of the San Andreas Fault is near the southernmost part of California, where the right lateral movement ends. Further south, Baja California is spreading away from the Mexico mainland in a series of rifting zones in the Sea of Cortez (see the map on page 30.) The area west of the spreading zone is also part of the Pacific Plate captured by translational movement to the north. The Pacific Plate boundary extends more than 11,000 miles along the North and South American continents.

On the north end at Cape Mendocino, the San Andreas Fault merges into a triple junction with the Gorda Plate as it joins the Pacific Plate and the North American Plate. North of the triple junction, the spreading zones cause the eastern section of the Pacific Plate to plunge under Oregon and Washington.

The San Andreas Fault came into existence from 23–29 million years ago, after the pre-existing Farallon Plate was consumed under the North American Plate. Prior to this time, the spreading centers were in a tensional mode and offshore basins were formed along much of the coast. Subsequently, the movement along the San Andreas Fault has carried these basins to the north. Some Miocene rocks in the Gualala Basin are thought to have formed as far as 350 miles south of their current position.

The strike-slip portion of the SAF is 838 miles in length and stretches from Cape Mendocino to the Southern California/Mexico border. In 1906, over 270 miles of the fault moved 8'– 22' in a few seconds, or at most minutes. Other movements of the SAF occurred over shorter segments as the pressure along the fault was released. It may take hundreds of years or even longer for the entire 838 miles to move and adjust. Earthquakes at a specific location along the fault occur in 100 to 300 year intervals. Thus, here in Northern California, we may be due at any time—or maybe not for another 200 years.

Many studies have been undertaken to determine how much slip has occurred. The current thinking is around 195 miles of total displacement. The Pinnacles National Monument south of Salinas seems to match volcanic rocks in the Mojave Desert with 192 miles of displacement. Some Miocene sandstone beds found south of San Francisco seem to match sandstones 225 miles further south, plus located on the opposite side of the fault. If we assume the average movement along the San Andreas Fault was one-half inch per year, this calculates as 4.2' per 100 years, which translates as 7.95 miles per one million years. For 26 million years, we calculate 206 miles of movement; for 29 million years, we calculate 231 miles of movement.

We speak of the San Andreas Fault as if it is a single trace, but in reality it is a zone up to one-half mile in width. Successive movements may shift from one trace to another. The epicenters of seismic events on the fault are five or six miles or more in depth near the base of the continental crust. The surface trace of the fault in a specific movement may not always reach the surface at the same place. In the millions of years of plate movements along the San Andreas Fault, the surface trace has shifted from offshore locations to land locations in a zone over 10 miles in width.

Other regional faults run parallel to or merge with the SAF and also exhibit strike slip movement. The San Gregorio Fault south of San Francisco merges with the SAF at Pt. Reyes. The Calaveras, the Hayward-Rodgers Creek, and the Healdsburg Faults are east of the SAF and are also considered active faults with strike slip movements. All of these faults have experienced at least several miles of offset movement. Seismic work in the offshore basins indicates several offshore parallel faults with suggested strike slip movements. Looking at all these faults as a package gives a zone of nearly 100 miles in width which is under the stress of or is affected by these two grinding plates.

In the area of this study, the San Andreas Fault Zone basically follows the trace or west side of the mainstem of the Gualala River and the Garcia River. An older trace of the SAF occurs along the ridge top on The Sea Ranch and east of Gualala near Bower Park. This trace has been called the Gualala Fault or the Gualala Ridge Fault. The surface trace has been obliterated by weathering and erosion, but a series of sag ponds and depressions delineate the trace of the fault. In the Appendix Road Log section, I identify fault features you may see at specific locations from Bodega Bay where the fault comes onshore and then offshore, and where it again comes onshore just south of Fort Ross, and finally to Alder Creek, just north of Manchester, where the SAF moves under the Pacific Ocean.

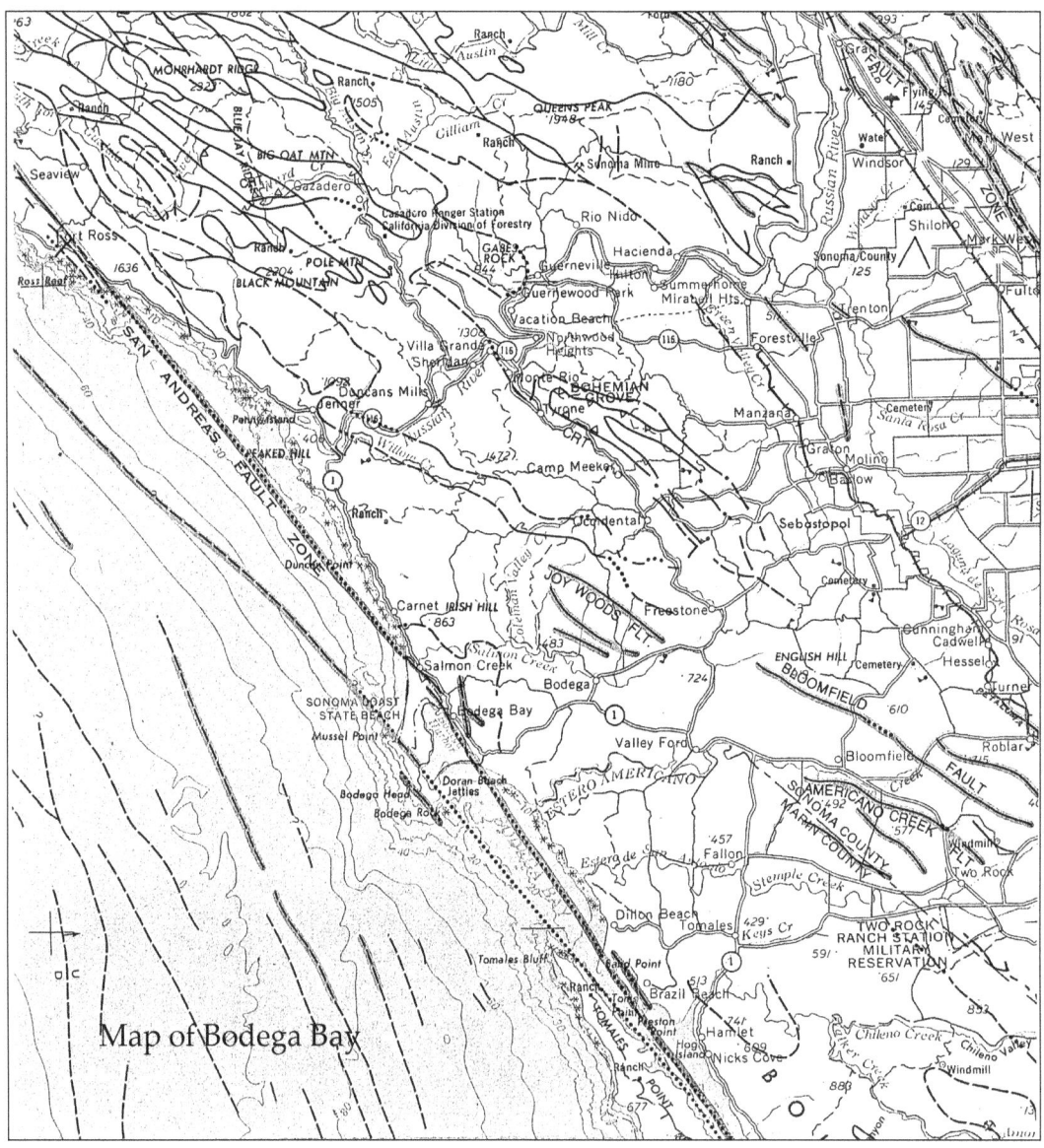

Recent Fault Movements since the beginning of the Pleistocene. Area shown is part of Santa Rosa Quadrangle. Heavy lines depict recent movements, the rest of the faults are older in origin, and many are related to folding of the Coastal Ranges. Source: California Department of Mines and Geology. 1982

Bodega Bay

Are Lucas Wharf & Inn At The Tides built on the fault?

The San Andreas Fault comes back onshore briefly as it cuts across the Bodega Head. It is located beneath the ocean to the south and cuts Point Reyes, forming Tomales Bay. Excellent guidebooks are available to lead you around Point Reyes both by auto and hiking trails. Larsen (1908) documented many of the features produced in the 1906 Earthquake.

Bodega Bay is less well-documented as to the damage produced by the 1906 San Andreas Fault movement. The picture is complicated by the fact that much of the displacement was under the east side of the bay. The trace then moves northward through a sand dune area, which easily obscures the actual fault trace(s).

An examination of the geologic map gives one an idea of the complexity of the area. The major bedrock formation east of the San Andreas Fault is the Franciscan Formation of Jurassic to Cretaceous age. The Franciscan has been faulted, folded, and uplifted in the coastal ranges. Sea level changes have left coastal deposits of sand and gravel on the marine terraces cut during Interglacial times. Weathering of the bedrock also obscures the numerous faults which are shown on the map as dashed lines indicating their approximate trace.

Of note is the Salmon Creek Fault Zone which is nearly one-half mile in width. The rocks within the zone are highly brecciated. The zone trends northwest and goes offshore at Portuguese Beach. It appears this is an old fault associated with the folding of the Franciscan Formation, and did not move during the 1906 Earthquake.

The rocks on Bodega Head are granite and granodiorite, and match the granodiorite located eight miles south on the Point Reyes peninsula. These rocks are thought to be 100 million years (100ma) in age. Bodega Head is bounded on the east (which is the west side of the bay) by the San Gregorio Fault, which also has experienced considerable lateral movement. Estimates of movement along the fault there are around 90 - 100 miles. The granites west of the San Gregorio Fault are part of the Salinian Block which are matched to rocks found 300 miles to the south in the Salinas Valley.

The geologic literature is not consistent in that some authors consider everything west of the San Andreas Fault as being in the Salinian block. Others refer to the Gualala Block as the area west of the San Andreas Fault. I confine the use of the term Salinian Block to refer to rocks located west of the San Gregorio Fault, and refer to the zone in-between the two faults as the Gualala Block. Some people assert the San Gregorio joins the San Andreas as it crosses the Point Reyes Peninsula, while other maps show the San Gregorio trending northwest after crossing Bodega Bay.

Seismic maps of the offshore area indicate many faults which have been considered as adjustment faults; these occur as the offshore basins settled and filled with sediments. If several of these faults have large amounts of lateral movement, then we need to reconsider the width of the plate boundary between the North American and Pacific Plates. The distance between the Healdsburg Fault and the offshore San Gregorio Fault is approximately 30 miles. Minor movements may also have occurred both east and west of these faults, thus extending the affected area to 40 or 50 miles.

Bodega Bay is approximately one mile wide, which is also the considered width of the San Andreas Fault zone. In the 1906 event, the movement is thought to have been in the same range as the 20' experienced at Point Reyes. It was reported that the fault trace was near the eastern shore of the bay, and that 18" of vertical movement occurred on another trace in the dunes. The sag ponds and pressure ridges formed were subsequently erased by wind action in the sand dunes.

Fort Ross Area

Did the Russians feel any earthquakes?

The Fort Ross area is interesting from a historical perspective as well as from a geologic perspective. The Russians built and operated the fort from 1812-1841 and reportedly took 600,000 sea otter pelts from our coastal region. Over 170 years later, the local sea otter population has yet to recover. Maybe this tells us something about the fragility of an environment and the effect of humans on that environment.

In 1906, the San Francisco Earthquake destroyed many of Fort Ross's buildings. The current group of buildings are all reconstructions of several of the original buildings. When the small group of Russian trappers were active at the fort, they hired many Indians who actually did most of the work in killing and processing the sea otters.

The San Andreas Fault goes out to sea at the north end of Bodega Bay and comes back onshore south of Fort Ross at milepost SON 30.30 just to the south of Mill Creek. As you travel north on Highway One, the rocks change from Franciscan sandstones to lighter-colored Cretaceous rocks on the steep bluff slope to the ocean. North of the Caltrans concrete faux-stone wall, Terrace I appears at the fault trace, and then widens to the north at Pedotti Ranch. *(Although only one trace of the San Andreas Fault has been identified at this point, I suspect another trace exists at or near the Caltrans wall.)* The United States Geological Survey (USGS) has trenched across the fault near Mill Creek to determine the rate of movement, amount of fault offset, etc. (Simpson 1996, Prentice 2000).

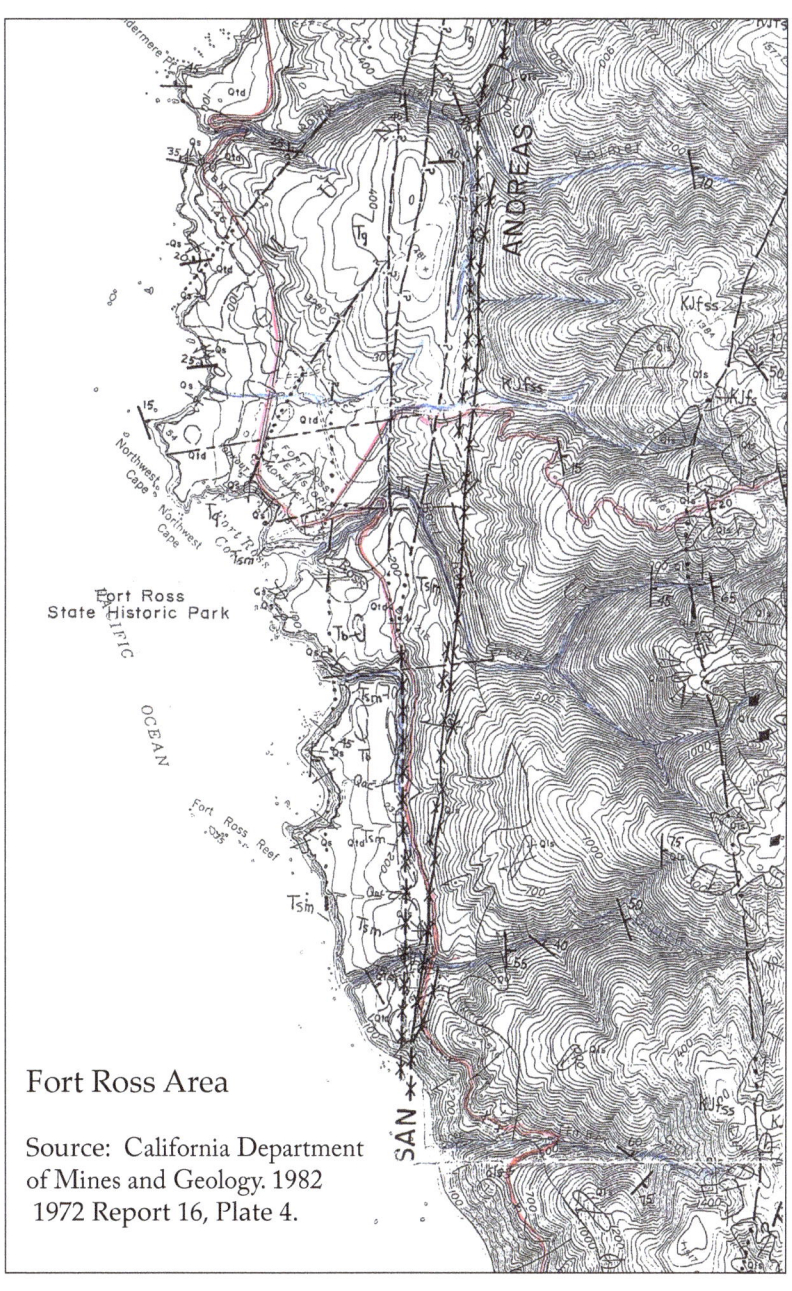

Fort Ross Area

Source: California Department of Mines and Geology. 1982
1972 Report 16, Plate 4.

Note the offsets of the streams, Kolmer Gulch, Fort Ross Creek, and Mill Creek. No offset on Timber Gulch. Highway 1 follows the SAF from the faux rock wall to north of Fort Ross

Proceeding northward near Fort Ross, the fault trace splits into several traces. The hillside to the east is a massive collection of landslides, slump, and creep features. The gully on the west side of the highway is a fault segment. The recent highway repair at SON 33.00 encountered fault gouge and slump materials at the north end. The Russians reportedly mined coal along the

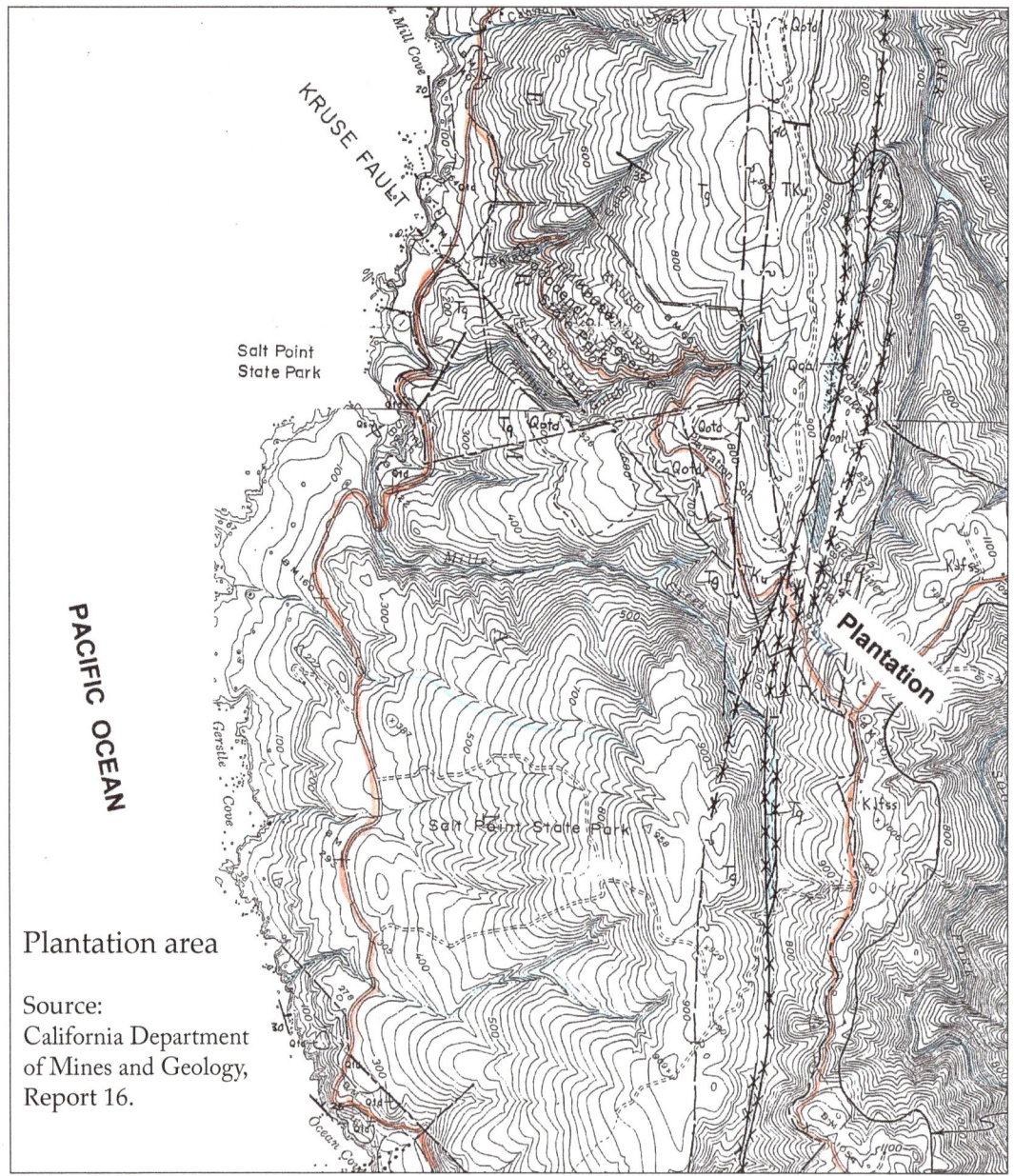

The six cracks in the 1906 event are shown here as two or three major traces. The most obvious physical features marking the San Andreas Fault (SAF) are the sag ponds and the valleys which follow the fault traces.

creek adjacent south of the Fort. We have not been able to find any traces of coal in the stream. It is possible the deposit was very small and caused by pressure from fault movement. The underlying rocks are marine deposits and rarely contain any coal.

Taking Fort Ross Road uphill one-half mile to the east, there are four fault traces that cross the road. Sag ponds and marshes occur north along the farm road east of the old barns. Just east of the sag ponds is a second low area with very strange trees, apparently disrupted by creep along the fault. The trees branch, twist, and fork in odd and mysterious forms. *(In my mind's eye, I see the scene from a Disney movie where the Sorcerer's Apprentice is being chased by the trees.)*

Plantation

Will the next quake also miss all the buildings?

Plantation is situated in a small valley at 742' elevation just west of the South Fork of the Gualala River. The property is a small sheep, cattle, and horse ranch operating as a children's summer camp. To get there, turn east at mile marker SON 42.70 and drive through the Kruse Rhododendron Preserve for approximately 3.4 miles, which is only 1.5 miles from the coast. You may also reach Plantation by travelling north on Meyers Grade Road and Seaview Road for several miles along the coastal ridge.

It became clear in the great 1906 San Francisco Earthquake that Plantation sits on the San Andreas Fault. Six parallel cracks developed in a 270' wide zone which encompassed several of its few existing buildings. Apparently only one crack went through a building—a small cottage which suffered some damage, but was later rebuilt.

Tree tops were snapped by the severe shaking, and one tree at the east side of the property on the north side of the road was split by a fault trace. The fault traces had a strike of N. 38° W. and exhibited 1"- 6" in the road to 1' on the major trace of vertical movement. The upside displacement was on the west side of the faults. Horizontal movement in this section of the San Andreas Fault had a cumulative movement in the six cracks of 10' - 15', with the western side moving to the north.

Sag ponds are a common occurrence along the fault. Two large sag ponds were developed or rejuvenated by the 1906 movement. These occur in a band to the northwest of the property and are named on the 7.5 Minute Plantation Quadrangle as Lake Oliver and Lower Lake. Each of these ponds/marshes measure approximately 2000' in length. Several smaller sag ponds also developed in the 1906 event. Two small sag ponds were developed in the front yard of the main farmhouse and have been subsequently filled in. Agriculture and erosion have softened the land surface, so without the enclosed maps, it is difficult to trace many of the quake-related features.

On the east side of Lake Oliver is a pressure ridge 10' - 15' in height, which was pushed up by the grinding of the two plates. Two small sag ponds filled with pond lilies occur just east of the pressure ridge and identify one of the other fault traces.

Erosion occurs more easily along fault traces than in solid, unfractured rock. The surface streams in the area are adjusted to the fault traces. The South Fork of the Gualala River basically follows

Plantation, CA

Not only were the streams offset along the various traces of the San Andreas Fault, but several sag ponds were created. This is the largest sag pond in the area, and holds enough water for swimming in the summer. Two or three of the smallest sag ponds have been filled in, particularly those near the main house. The six cracks produced during the 1906 event have been filled in by erosion and deposition, and the activity of farm animals. Several redwood trees in the area evidence snapped tops, and a scar exists on one redwood tree that was split by the fault. You can see it next to the road on the east side of the property.

the fault trend. Tributary streams generally flow westward into the Pacific Ocean, or eastward into the South Fork of the Gualala River. When the headward reaches of these streams encounter the fault zones, they are turned by the moving fault and follow the fault traces. Miller Creek, just to the south of the Ranch, turns abruptly south following the fault for approximately .6 miles of offset. Northwest of the Ranch, the Phillips Creek headwaters part of the stream abruptly turns northwest and southeast along the fault, with an apparent offset of .65 miles.

Stream offsets from south of Fort Ross to Plantation

Wow! Think of how many movements it must take to get one-half mile of stream offset!

The San Andreas Fault returns to land just south of Fort Ross at milepost SON 31.76. For 10 miles along the fault zone to Lake Oliver at Plantation, all the streams which cross the fault zone suddenly turn north along the fault zone, and then shift west into the ocean. These streams evidence repeated movements along the 1906 traces of the fault. The following nine streams from south to north are listed below with their measured offsets as indicated on the Fort Ross and Plantation 7.5 minute topographic map quadrangles.

1. Timber Gulch SON 29.90 to Mill Creek SON 30.75   .6 miles of offset

    San Andreas Fault comes onshore SON 30.30

2. Mill Creek SON 30.75 East and west of Hwy One   .1 mile of offset

3. Unnamed creek at Pedotti Ranch beach access runs parallel to Hwy One SON 31.00   .4 miles of offset

Entrance to campground SON 31.37

Fort Ross Entrance SON 33.00

4. Fort Ross Creek SON 32.45   .55 miles of offset

5. Kolmer Creek Gulch SON 34.36   .3 miles of offset

6. Timber Cove Creek SON 35.44   .5 miles of offset

7. Stockhoff Creek SON 37.24   .5 miles of offset

Stillwater Cove

8. Miller Creek SON 41.00   .6 miles of offset

9. Philips Gulch at Plantation   .65 miles offset

All of these streams were apparently consequent streams flowing westward off the coastal ridge into the ocean. The streams obviously predated the movements along the San Andreas Fault in this area. The rate of movement of the Pacific Plate is estimated at one-half mile in 125,000 years (at ½" per year). The maximum cold period of the last Glaciation was approximately 25,000 years ago, with the ocean sea level approximately 300' below the present sea level. During glacial times, the coastal area experienced much greater rainfall with a consequent increase in the amount of erosion. Local streams cut as much as 200' into the continental shelf near the mouths of the major streams.

Continuing north of this area, the San Andreas is further inland from the ocean and located behind the first coastal ridge. Streams to the north tend to follow the San Andreas and do not show obvious offsets like the above-listed streams. The major trace of the fault is along the mainstems of the Gualala River and the Garcia River.

Other structural offsets

The **Gualala Ridge Fault**, which is an older trace of the San Andreas, is located near the crest of the first ridge of coastal hills, and does not seem to show any significant stream offsets. This is either because its location is near the ridge top which had no cross-cutting streams, or because the age of the movements was so old that erosion has erased the evidence. As it cuts across the structural grain to its mouth, the three large bends in the Gualala River appear to be influenced by faulting. However the offsets are in hundreds of feet, not miles. Therefore the predominant movement along the San Andreas was along the mainstem faults near the 1906 trace, and not along the Gualala Ridge Fault. The bends in the lower section of the Gualala River, west of the San Andreas Fault, indicate the river existed before this coastal segment was uplifted.

Since the estuary segment of the Gualala River is on the west side of the San Andreas Fault, it cannot always have been connected to the Gualala River watershed. At the current rate of offset movement, the western land section (Pacific Plate) is moving 7.95 miles in one million years. The Wheatfield Fork is located 17.5 miles south of the Gualala River Estuary segment and seems to be at the axis of the reconstructed basin in which the Ohlson Ranch Formation sediments were deposited. This offset of 17.5 miles would place the mouth of the Gualala River at the Wheatfield Fork 2.2 million years ago. Point Arena (Alder Creek) is 20 miles north of Gualala and would have been opposite the Wheatfield Fork 2.5 million years ago. Beyond this 2.5 million year age, continued movement would make the mouth of the Wheatfield Fork at the ocean.

Continuing this theory, the Gualala River Estuary was 30 miles further south to the mouth of the Russian River 3.8 million years ago. And ironically, the meandering Gualala Estuary looks like the meandering lower part of the Russian River.

Returning the discussion to The Sea Ranch, we find there are other offsets to consider. There is an offset of the Black Point Spilite along the Gualala Ridge Fault in the magnitude of one mile. Estimated at 600,000 years in age, thick and apparently undisturbed soil has developed over the fault on Terrace V, which would indicate plate movement had subsequently shifted to other

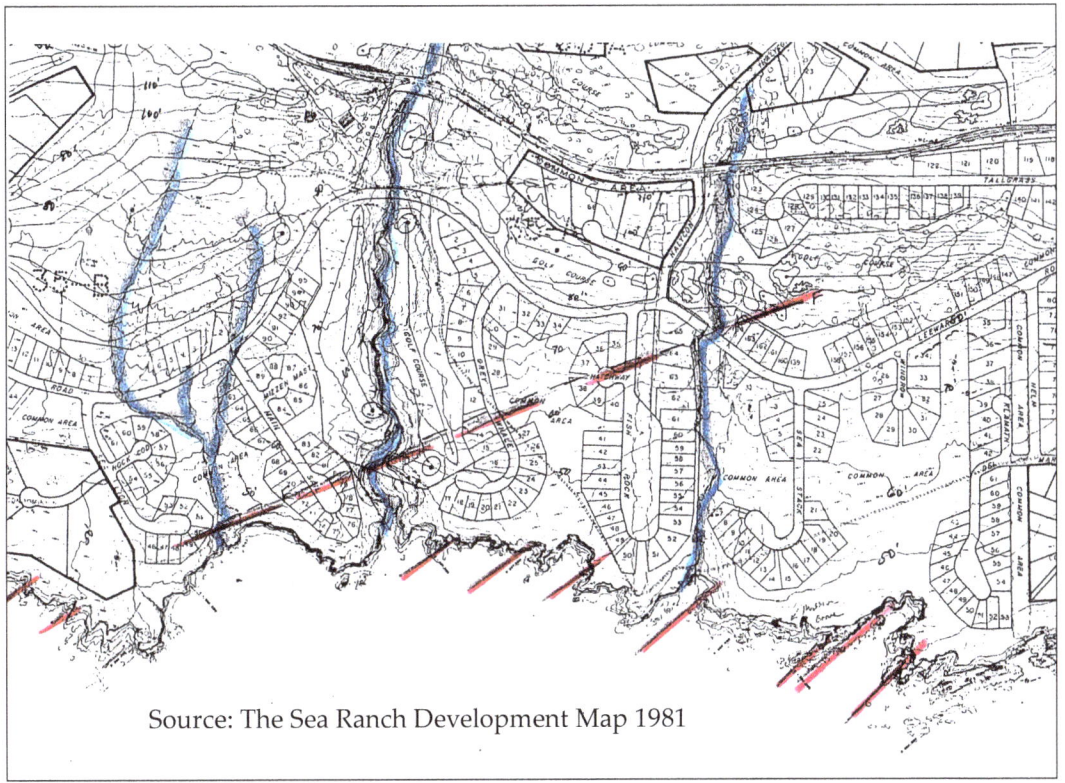

Source: The Sea Ranch Development Map 1981

Stream offsets at the north end of The Sea Ranch. Numerous faults are encountered in the bluff face all along our coast. The majority of these cracks are thousands or even millions of years old. They are related to the folding and faulting of these rocks as they have been uplifted and wrenched north along the San Andreas Fault (SAF). The stream offsets occur on the lower coastal terrace that was formed 81,000 years BP. We would expect the subsequently developed streams to flow downslope toward the ocean in a fairly straight path. It is interesting that these three streams show a similar amount of offset. I therefore suggest a possible fault as the cause of the offset. Other explanations are possible in that the coastal terrace deposits change in texture from sand to gravel. Sometimes bedrock knobs protrude above or near the surface and may deflect the stream paths. A few years ago, we experienced a small earthquake, with the epicenter just offshore from this area in the deep ocean basin. It is probably not related to this feature, but indicates there are offshore active faults, as well as the onshore San Andreas Fault.

fault traces. An offshore fault (actually several faults) have been identified from seismic data, and one is referenced as the Gualala Fault. This fault appears to merge with the San Andreas Fault at Bodega Head. A magnetic anomaly occurs just west of this fault near the center of the Gualala Basin. This would be about five miles offshore of the north end of The Sea Ranch. *I interpret this anomaly as being related to the Black Point Spilite as it is of similar magnetic magnitude allowing for sediment cover. The Black Point Spilite appears to be offset a mile or more along this fault.*

At the north end of The Sea Ranch on the lower terrace (Terrace I) which was formed approximately 81,000 yrs. ago during an Interglacial high stand of the ocean, there are some minor offsets of coastal streams crossing the terrace. Salal Creek and Halycon Creek, located 1700' apart, each have a jog in their stream trace of 100' – 150'. If this is truly related to an offset along a fault, and not the coincidence of two adjacent streams meandering in the same direction, then several movements have occurred on the terrace in the last 81,000 yrs.

Mapping the Black Point Spilite (Kbs)

Looks like lava to me, why do they call it a spilite?

The *Black Point Spilite* is the oldest rock found west of the San Andreas Fault on The Sea Ranch—deposited 80-160 million years ago. It has a long, complicated history of deposition offshore in an island arc, uplift and erosion, subsidence and burial, subduction partially under the continental plate, uplift and erosion, thrust faulting, and successive movement along the San Andreas Fault of several hundred miles into our area. The shifting of the surface trace and position of the San Andreas Fault, from one event to another, have split the formation into several blocks.

The Black Point Spilite outcrops along the bluff edge from just north of Black Point and extends to the north end of Pebble Beach, a distance of more than 8000'. Coastal Terrace Deposits of sand and gravel cover the surface of the Black Point Spilite on the lower two coastal terraces, which are known locally as The Sea Ranch Meadows. In Unit 7, north of Galleons Reach, fossil sea stacks of Black Point Spilite poke up through the surface of the Sea Ranch Meadows. Outcrops also occur just east of Highway One from highway markers SON 51.64 to SON 51.75.

East of Highway One, the Black Point Spilite extends upslope under a blanket of soil and weathered bedrock, becoming more difficult to trace and map. It is probably bounded on the east of The Sea Ranch by the Gualala Ridge Fault, which is an older trace of the San Andreas Fault.

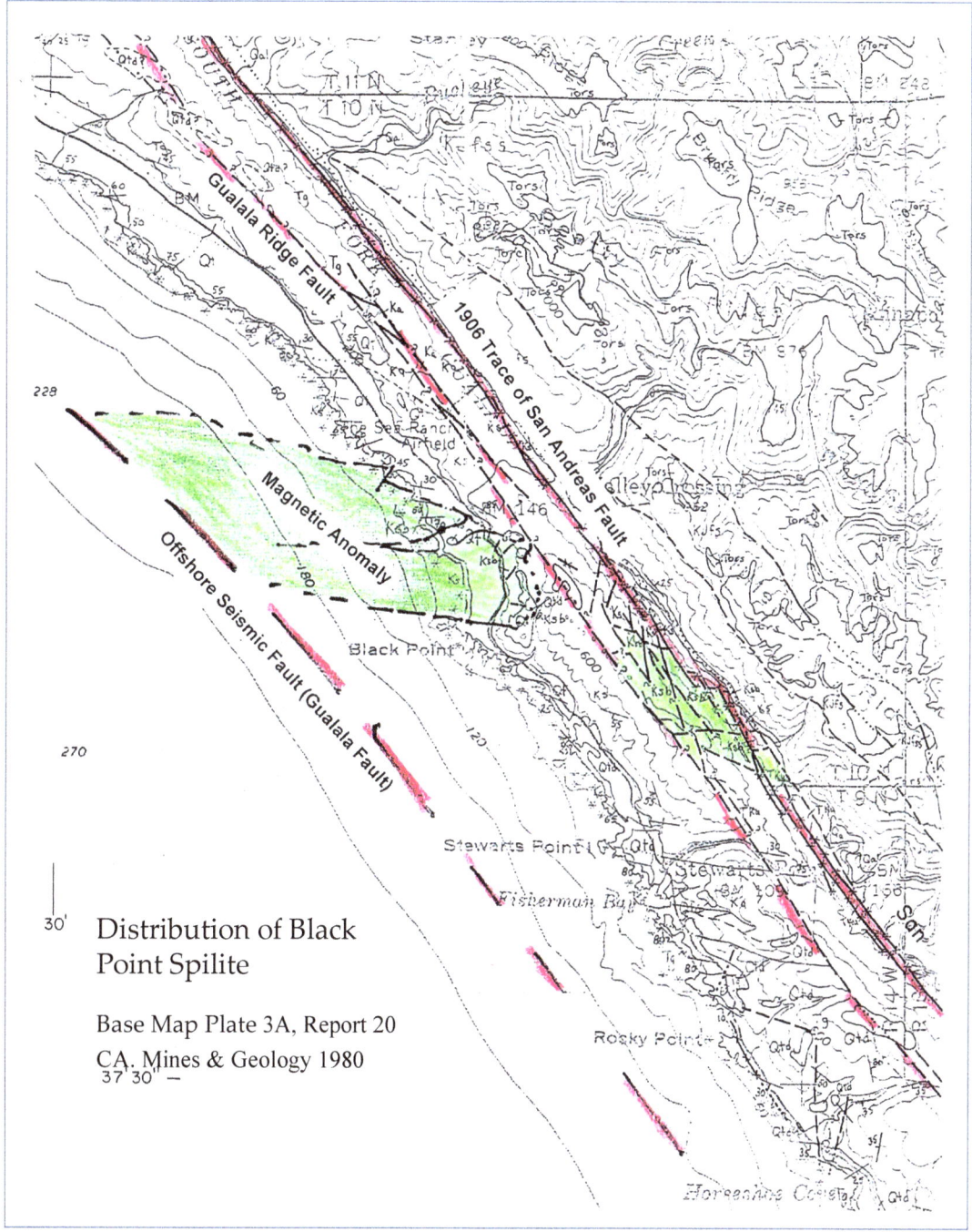

Black Point Spilite is the oldest rock present in the area west of the San Andreas Fault. Younger rocks occur both north and south identifying the feature as an anticline. Strike slip movement offsets the formation on the Gualala Ridge Fault and offshore on the seismic identified Gualala Fault. At the north end of The Sea Ranch another magnetic anomaly occurs several miles offshore, which is probably an indication of the formation at depth in the offshore basin.

Published maps indicate the Black Point Spilite outcrops south of The Sea Ranch at the top of the ridge of the Richardson Ranch, north of Skaggs Spring Road. (In personal reconnaissance of the small streams draining westward to the ocean, I have not encountered any substantial erosional fragments of Black Point Spilite. However, in reconnaissance two miles south of the twin bridges on Annapolis Road along the South Fork of the Gualala River, I located two large boulders of Black Point Spilite at river's edge that appear to be bedrock in place. If these boulders are not in place, an alternative explanation for the outcrop is that it may have been part of an adjacent landslide produced by the 1906 San Andreas earthquake event.) Fault breccia occurs on the east side of the stream approximately 50' from the Black Point Spilite boulders. The bedrock on the east bank of the stream is Franciscan sandstone. A sag pond is situated several hundred feet north on the flood plain bank east of the river. (Both the breccia and the sag pond are in line, and I interpret these features to have been formed during the 1906 San Andreas Fault movement event.)

In 1944, Charles E. Weaver published the first major examination of this coastal area from Fort Ross to Point Arena, identifying the Black Point Anticline with the oldest rocks at the center and younger rocks both north and south along the coast. Weaver grouped the oldest rocks into the Gualala Formation, with the igneous rock at the core identified as basalt and diabase of Franciscan Age. Later, it was renamed and identified as the Black Point Spilite. The contact on the south end of Black Point Beach is indicated as a fault contact of breccia and conglomerate. The fault is easily identifiable as a zone approximately 20' in thickness with brown to grey conglomerate on the south side, and highly fractured (brecciated) black basaltic rock on the north side. The fault trace dips 40° to the southeast.

The Black Point Spilite is very hard, dark green to black in color, and highly faulted and fractured. It is a very fine crystalline, basaltic, igneous rock, sometimes containing larger crystals (phenocrysts) of quartz, calcite, and pyrite, etc. Some zones exhibit small vesicles of gaseous inclusions formed during magmatic extrusion. Some zones show remnant features of pillow lava formed when hot lava flows into water. The beach sand along Black Point Beach is green to black in color, but is mixed with lighter-colored sand, eroded from the overlying adjacent formations. The black sand is often sorted by the waves into wavy lines on the beach. Some of it is magnetic and easily separated by using a magnet. Magnetic surveys indicate strong magnetic anomalies over the Black Point Spilite as compared to the adjacent rock formations or other igneous rocks such as the Bodega Bay Granite or the Iversen Basalt.

The Black Point Spilite is thought to have been deposited as an underwater series of basalt lava flows in a back arc position from an Island Arc series of volcanoes. (The Japanese islands are a volcanic arc series with the back arc position to the west of Japan). The site of these lava flows was hundreds of miles to the south and occurred 80–160 million years ago. *(This age may be inaccurate due to the later metamorphism that could have affected the radiometric measurements.)*

After deposition, as the California Coastal Ranges were folded and faulted, these lava rocks were subducted underneath the North American continent and metamorphosed by heat and pressure to the low grade *greenschist facies*. Subsequent to the subduction, the rocks were uplifted and partially eroded, and then covered with younger unmetamorphosed sediments. And, finally, in the last 15+ million years of plate movement along the San Andreas Fault, these rocks have been carried north to our area.

The current structural attitude of the Black Point Spilite is an east-west trending anticline located on the south end of The Sea Ranch. The adjacent overlying rocks get younger in both directions to the north and south. The immediate overlying rocks are conglomerates and sandstones belonging to the Stewarts Point Member of the Gualala Formation of late Cretaceous age which were deposited 70–80 million years ago. Overlying these rocks is the younger Anchor Bay Member of the Gualala Formation, again consisting of sandstones, mudstones, and conglomerates. The Gualala Formation has been measured as over 1500 meters (over 4500') in thickness. The two members are differentiated from each other by slight differences in the composition of the sandstones and the conglomerates. The conglomerates in the Stewart's Point Member have cobbles of granite and quartzite, whereas the Anchor Bay Member has darker cobbles of basaltic rocks. Known as the German Rancho Formation of Paleocene and early to middle Eocene age, further north and south, on the flanks of the anticline, a younger formation occurs in a section of up to 6000 meters (over 18,000') in thickness.

All the formation contacts with the Black Point Spilite are faulted contacts, so we do not know what sized section of the Stewart's Point Member may be missing, or if there were older rocks deposited over the Black Point Spilite which have eroded or faulted out. According to previous studies, the basal conglomerates in the Stewarts Point Member do not contain any cobbles of Black Point Spilite. *(However, near the fault contact on the south end of Black Point beach, I have identified a thin section in the conglomerate that contains cobbles of Black Point Spilite.)*

The movement of the two continental plates—the North American Plate and the Pacific Plate—have fractured and faulted the rocks of the Gualala Basin, but have impacted the rocks adjacent to the San Andreas Fault Zone most intensely. We tend to focus on the 1906 trace of the fault as we have the most substantial plus relatively recent evidence of its location. However, the San Andreas is a zone and not a single trace. The Gualala and Garcia Rivers have fairly wide valleys formed along numerous old traces of the fault zone. Offshore faults have been identified by seismic exploration and periodically experience fault tremors. Fault traces occur in the rocks along the bluff and along the beaches which run parallel to the San Andreas Fault; however most of these faults only exhibit minor offset of rock bedding.

Running along the top of the hills on The Sea Ranch and through the Gualala area is an old trace of the San Andreas Fault. It is difficult to map the exact trace because erosion has destroyed many of the old fault features. However sags and depressions, pressure ridges, and sag ponds do occur in numerous spots along the trace. The Black Point Spilite is offset by the Gualala Ridge Fault by 6000'–7000', which indicates this trace was active for a considerable length of time. Consider: if the offset of 6000' occurred in 10' of displacement movements each 200 years, then 120,000 years of time is required for the offset. We don't know when the last movement on this trace occurred, but it probably happened thousands of years ago, yet not so long ago as to allow erosion to erase the numerous seismic features still in existence. The sag ponds at Lumberjack Close and near The Sea Ranch Association office appear nearly as fresh as the 1906 sag ponds along the Earthquake Trail above the Hot Spot.

On regional Aero-magnetic and gravity maps, the Black Point Spilite forms prominent anomalies. About six miles offshore from The Sea Ranch, a parallel fault to the San Andreas Fault has been mapped as one of several faults identified from seismic surveys. A large magnetic anomaly occurs on the west side of this fault. *(I interpret this anomaly as another deposit of Black Point Spilite covered by thousands of feet of sediments in the Gualala Basin.)* Additional magnetic anomalies occur north and west of Point Arena, and may also indicate the presence of Black Point Spilite at depth beneath the overlying sediments.

Point Arena

This quaint, sleepy-looking town is actually the hub of many local activities.

Point Arena is a small, California city with its own city government, schools, churches, a theater, restaurants, a motel under reconstruction, a fine bakery, art galleries, and more tourist-oriented delights. Arena Cove has a working pier and is the only active port with harbor facilities within our project area north of Bodega Bay, 65 miles to the south. It is approximately 40 miles north to Noyo Harbor and Fort Bragg.

The city is here because of its geologic setting on the Arena Cove. Incidentally, most cities and towns, were at least initially located on a feature controlled by the geology (i.e: coves, rivers, mountain passes, waterfalls, lakes, etc.). Many different geologic features are easily accessible in the area. Recently, the Stornetta lands, consisting of 1665 acres from Arena Cove to Manchester Beach, have been designated as a national monument and added offshore to the Greater Farallones National Marine Sanctuary.

The rocks are part of the Monterey formation and the Point Arena formation. These rocks have been faulted and folded into complex structures. On the south part of Arena Cove is a huge anticline. At low tide, you can walk to the center of the anticline. Along Harbor Road, the rock dips range from gently north-dipping beds to vertical beds. One of the problems geologists encounter in studying rock formations is determining the amount of real dip versus apparent dip in the rock formations. The problem is caused by outcrops which are cut obliquely across the rock strike. The outcrop behind the Arena Pharmacy Building will explain this concept to you. As you enter the parking lot north of the pharmacy, one encounters steeply north-dipping rocks behind the building. The excavated parking lot cuts the rocks and they wrap around to the west, looking like a syncline. Actually, the rocks are in the same plane and not folded into a syncline as first appears to be the case. Check it out on your next visit.

In Point Arena, the San Andreas Fault is located two miles east of town. There are however numerous other faults that control the topography of the area. During the 1906 Earthquake, considerable damage was done to many of the buildings in town. The Point Arena Lighthouse was destroyed during the event and was reconstructed afterwards as one of the first earthquake-reinforced concrete buildings structurally designed to withstand future earthquakes. The Lighthouse is owned by a local nonprofit group and is a marvelous vantage point for viewing the point as well as the adjacent Stornetta Lands.

Several adjustment faults cut the point and sea caves are developing, undercutting the land.

Aerial view of Point Arena Lighthouse, 2006. The sinkhole in the foreground has developed along a fault that cuts across the point. Currently erosion of the steep walls has opened the hole to basically a slot through the point. From the photo it is obvious that the whole point, including the lighthouse, is threatened by the ongoing erosion.

This photo BELOW, taken in 2015, shows how the sinkhole has now been lost to erosion.

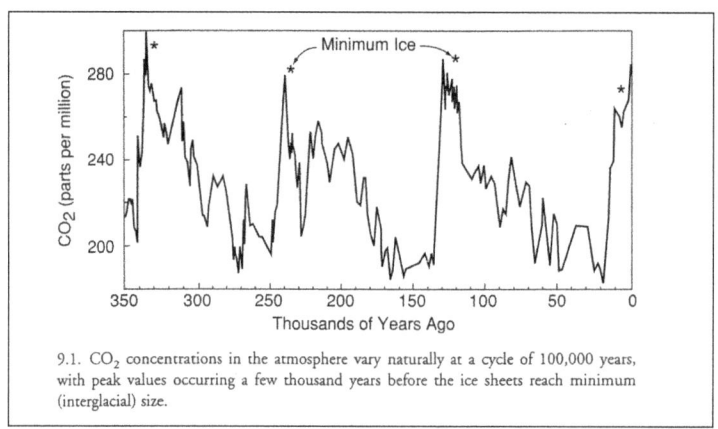

9.1. CO_2 concentrations in the atmosphere vary naturally at a cycle of 100,000 years, with peak values occurring a few thousand years before the ice sheets reach minimum (interglacial) size.

Many studies of deep sea cores, glacial cores in Antarctica and Greenland, temperature curves, and methane as well as carbon dioxide produce similar curves. This illustration shows the last three glacial periods and the warm interglacials. Note the 100,000-year cold cycles and the warm 20,000-year cycles. Of significance is the trend of glacial periods to become increasingly colder toward the end of their cycle, and the warm cycle is very abrupt in forming. Humans arrived in North America during the peak of the last ice age. How did they get here and how did they survive?

Source: Ruddiman, William F., *Plows, Plagues and Petroleum – How Humans Took Control of Climate* (Princeton University Press, 2005, page 85)

Pleistocene Glaciation: cause & development of coastal marine terraces

Chapter 5

You mean glaciers affected our area, but there were no glaciers here?

The California coastline exhibits as many as 14 benches or terraces cut into the coastal hills. These benches are thought to have been produced by wave action during the past 2.9 million years. Along the Sonoma-Mendocino coastline, we can identify four to six terraces from the ocean bluff to the top of the first line of coastal hills. Under the ocean surface, there are at least three more wave-cut terraces descending to more than 300' below sea level.

Several questions immediately arise. How were the terraces formed? When were they formed? How long a period of time does a terrace represent? What is the rate of erosion and bluff retreat? Are some terraces missing, or completely eroded? Since the terraces are now raised above sea level, how were they raised? What is the rate of uplift of the land? How do scientists address these questions?

Many of the explanations come from other places far away from our coast: from coral reefs in the South Pacific, ice cores from Greenland and Antarctica, Europe and the Midwest United States, and even the Sahara Desert. Before the mid-19th century, we had little scientific knowledge about glaciers, most of which related to mountain glaciers in the Swiss Alps. From studying *Alpine Glaciers*, especially the related glacial deposits and moraines, in 1837, Louis Agassiz proposed that glaciers had previously covered much of France and Northern Europe. By mid-century, the theory of continental glaciation and the Pleistocene Ice Age became accepted by scientists and the general public. In the 175+ years since, geologists have mapped glacial deposits wherever glaciers were present during the Pleistocene. Glacial deposits have also been identified in older lithified rocks dating back to the Proterozoic, over two billion years ago.

The Pleistocene Ice Ages began 2.58 million years ago and continue through present time into the future. The cycles of glaciation cold periods and interglacial warm periods range in 100,000-year glacial cycles and 20,000-year interglacial warm periods, with some smaller interruptions or variations. The last 400,000 years have been the most documented and we can see evidence on our landscapes of these cycles—even though we are hundreds or thousands of miles from the sites of large ice fields.

The continental glaciers were exceedingly dense, one to one and a half miles in thickness. The tremendous weight of the ice pushed the crustal rocks down into the mantle. Even today, 10,000 years after the last major melting of the ice, the land in northern Canada and northern Europe is rebounding from the loss of the ice weight! Glacial moraines, erratic boulders, and grooves and ridges mark the paths of glacial flow and define the parameters of original ice sheets.

The non-glaciated areas, including our local coast, were affected by changes in climatic patterns, the amount of rainfall, and by fluctuations of sea level. At glacial maximum times, the sea level dropped by as much as 300' which places the ocean area four to five miles off the current coast. Sharp canyons were cut into the adjacent coastal ranges

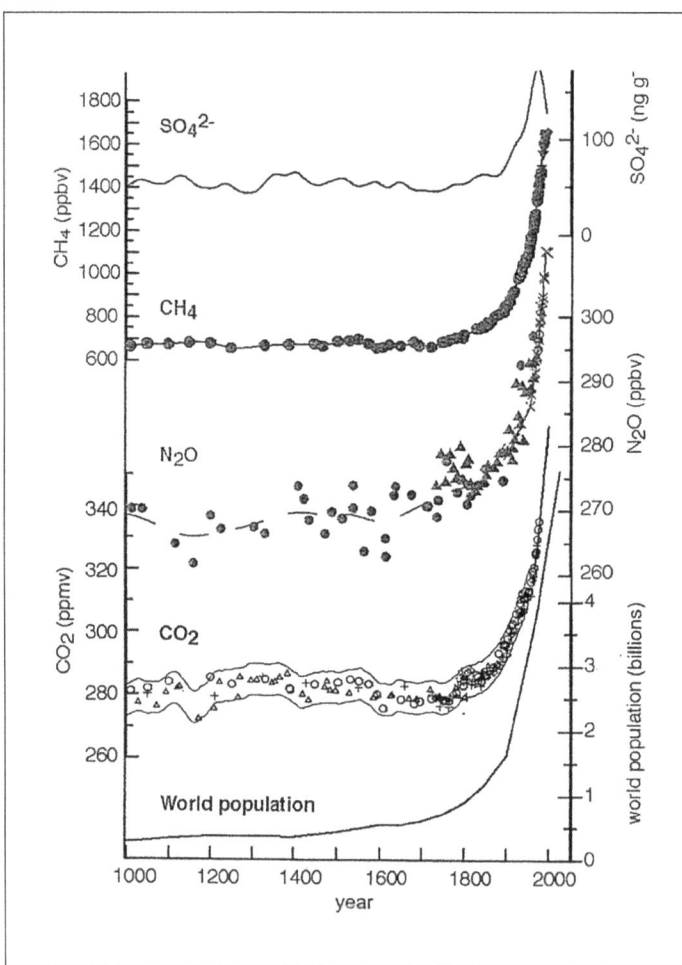

These are the so-called "Hockey Stick" graphs first produced in 1998 which were presented to Congress in 2000 and published in Al Gore's book and drew so much criticism. The 15 years since are bearing out the rapid changes shown on the graphs. The world population jumps after 1850 with the Industrial Revolution. Carbon dioxide, nitrous oxide and methane all show similar trends. Can we slow or stop these trends?

Source: Bradley, Raymond, *Global Warming and Political Intimidation - How Politicians Cracked Down on Scientists as the Earth Heated Up* (Univ. of Massachusetts Press, 2011)

and into the shelf area, which is now below sea level. The mouths of most of our major coastal rivers have buried valleys cut 200' below the current sea level.

The Pleistocene, 2.58 million years, is marked by 30 cycles of warm and 30 cycles of cold. The causes of continental glaciations are drawn from the confluence of several events. In 1941, Serbian scientist Milutin Milankovitch proposed there were three factors which combined to produce the climate changes necessary for continental glaciations:

1. **Eccentricity**. Earth's orbit around the sun is an ellipse which changes through time, in a period of about 100,000 years. Our planet is traveling at 67,000 m.p.h. and moves in its path from 91.4 million miles (Perihelion) to 94.5 million miles (Aphelion) from the sun.

2. **Tilt of Earth's axis**. The axis of our planet is currently 23.5° from the vertical. This tilt varies through time from 22.1° to 24.5° in a 40,000-year cycle. The northern hemisphere is more affected by this change because the continents are concentrated in the north. The southern hemisphere is mostly ocean, except for Antarctica, and the ocean dampens the temperature swings.

3. **Precession of the Earth's axis**. Our planet's axis slowly rotates around the pole over a period of about 20,000 years. If the northern hemisphere is facing the sun at the Perihelion (closest point in Earth's orbit), then warmer summers will melt the ice. When Earth's northern hemisphere summer occurs at the Aphelion (furthest point in the ellipse), then the summer is cooler and less melting occurs. When this happens, the winters are warmer and more snow may accumulate on the glaciers.

The combination of these factors seems to explain the waxing and waning of the continental glaciers. However, the ice cores, fossil evidence, sedimentary deposits, etc. indicate the cycles were initiated or terminated very rapidly—possibly in as short a time as 200 years. The 30 cold periods and 30 warm periods also show shorter cyclical changes from warm to cold or cold to warm. In historic terms, we saw the Vikings live in Greenland about 1000 CE which, shortly afterwards, was followed by the Little Ice Age in Europe, which ended about 1850. We have also seen very cold years caused by volcanic eruptions, i.e. Krakatoa 1883 and Tambora 1815. The warm Gulf Stream Current which bathes Northern Europe with warm water is currently being interrupted or affected by rapidly melting ice in Greenland. This could have a pronounced climatic effect and slow glacial melting—even plunging Europe into another "Little Ice Age."

Another factor to be considered is that the sun may not always radiate the same amount of heat. We used to believe the sun gave out consistent heat into the solar system. Our study of previous life on Earth has given us information on differing oxygen and carbon dioxide levels, different climates, and different periods of hot and cold, all of which have led to the conclusion of fluctuating heat output from the sun. Galileo first noticed that sunspots he saw through his telescope seemed to be cyclical in nature, varying from year to year. Large solar storms bombard our planet with cosmic rays which not only affect satellites and communication but could potentially affect the atmosphere, such as the ozone layer. These solar storms also produce the colorful Aurora Borealis in northern hemispheres and Aurora Australis in southern hemispheres.

Given all these factors, it is difficult to develop a realistic model which can predict when this period of climate change will end, and when the next continental glaciation will commence. Weather patterns in the short-term cannot be predicted more than a few weeks in advance. A slight shift in a storm path, a volcanic eruption on the other side of the world, a solar flare, or a variety of other factors can affect the outcome of any model. Scientists must keep collecting data and working on refining our models.

If we are only 10,000 years into a 20,000 year warm period, we might expect the glaciers in Greenland and Antarctica to nearly disappear, as they did in the last Interglacial Period. So the question is, should we build a boat for our grandchildren, or buy them skis and a fur jacket?

Comparison rates of uplift and elevations of California Coastal Terraces

How can adjacent terraces formed at the same time be found at different elevations?

Studies along the California coastline indicate there are different numbers of coastal terraces, ranging from two or three terraces to as many as 14 terraces, all in different areas. Obviously the sea level has changed many times. In some cases, a prolonged period of sea level rise has allowed erosion to completely obliterate a previous terrace, thus explaining why some areas have fewer terraces.

Concurrent with changing sea levels, there has been uplift of the land, which further complicates the correlation of the terraces. As the Pacific plate moves northward, it slowly buckles into ridges and saddles. Faults occasionally break the Gualala Block in an east-west direction, at an angle to or even perpendicular with the San Andreas Fault. These blocks are rising at different rates, further complicating the picture. The adjoining Coastal Ranges on the North American Plate are also being differentially uplifted.

Consequently, the unraveling of the terrace ages and the correlation of the terraces into adjoining areas is very difficult and complicated. The lower terrace, Terrace I of this report, is the most well-preserved and easiest to **use for comparison** to delineate and compare adjacent sections along the coast. Terraces are identified at their most landward edge, their highest end, as the lowermost parts of the terraces are being (or have been) differentially eroded by the current rise, or by a previous rise in sea level.

In her thesis, Carol Prentice (1989), now with the U.S. Geological Survey (USGS), identified the 83,000 years before present (83ka BP) terrace (Terrace I) at 85' elevation (her Area 1) and at 162' elevation (Area 3) north of Alder Creek in the Point Arena area. The trace of the San Andreas Fault leaves the land at Alder Creek and separates these two areas.

On most of The Sea Ranch, Terrace I occurs at 80' elevation. Near Fort Ross, Terrace I occurs at 98' elevation. On the Monterey Peninsula, Bradley & Griggs (1976) identified the lower terrace at 33' elevation with an age of 103ka. *(I suggest this terrace to be of the 83ka age.)* Hayes & Michel (2010) place the youngest terrace at Bolinas at 180' elevation, at Waddell Bluff at 150' elevation, and northwest of Santa Cruz at 140' elevation.

Gently sloping Terrace I. south of Point Arena Lighthouse

Sea level high stands which cut terraces are thought to have occurred at the following times: 83ka, 103ka, 120ka, 152ka, 202ka, 214ka, 305ka, 320ka, 336ka, 430ka, 740ka, 880ka, 1100ka. *(No doubt additional terraces were cut during sea level high stands back to the beginning of the Pleistocene (2580ka) 2,580,000 years ago).* Terrace I, formed 80–83ka years ago, is the most well-preserved and easiest to study. It was formed by wave erosion as the seas rose after the melting of the continental glaciers, known as the Illinoisan Glaciation (Ermian in Europe.) When the most recent Wisconsinan Glaciation began, the seas retreated, leaving beach sands and gravels covering the wave-cut terrace. The sea level dropped by 300' or more making the position of the ocean four or five miles off the current coastline. Major coastal streams cut down into the hills and the coastal terrace by as much as 200'. Several of the coastal rivers have gravel fills near their current mouths of 180'–200', attesting to the much-lower sea level.

The Wisconsinan Glaciation peaked about 20ka –25ka. The glaciers rapidly began melting about 18,000 years before present (18ka), continuing to about 6,000 (6ka) years ago, causing a sea level rise of about 250'. Since 6ka, glacial melting has slowed with some short warm and cold periods, and sea level has risen about 50', which is about $1/10^{th}$ of an inch per year. Currently, sea levels appear to be rising at a more rapid rate.

Terrace I has an onshore width ranging from 0–4000', averaging about 2000'. Offshore, the ocean depth changes abruptly to about 60' in depth. *(I interpret this to be about the edge of Terrace I, before it was eroded by the current sea level rise.)* This depth is 1000' to 2000' offshore, and averages nearer 1000' offshore. Bluff erosion from the advancing ocean averages two to three inches per year. Thus, 6,000 years of erosion amounts to 1000' – 1500' of bluff retreat.

If Terrace I was originally 3000' in width, and cut at the rate of two or three inches per year, the total terrace width would have been cut in 12,000–18,000 years, which is nearly the average length of an Interglacial Period.

Coastal terraces along Sonoma & Mendocino coast

How did the ocean cut a terrace 2000' above sea level?

Some of the most prominent geological features along the coast are gently sloping coastal terraces. Along the Sonoma-Mendocino coast, we can identify five or six topographic benches or terraces. The lower terrace occurs next to the ocean at elevations of 30'–80' where it occurs west of the San Andreas Fault, and at nearly 200' elevation north of Alder Creek, east of the San Andreas Fault. The terrace varies in width from 0 – 4,000'. Along the Jenner Grade, Terrace I abruptly ends just south of the Pedotti Ranch buildings.

These terraces in our area were formed in the last 600,000+ years and mark different sea levels caused by the oceans rising and falling in response to Pleistocene Glaciation. As the glaciers melted, the sea level rose and waves cut terraces into the bedrock. As the glaciers re-formed and took water out of the oceans, the sea level fell. As the seas withdrew from the land, sand and gravel beaches were left on the wave-cut terraces.

On wave-cut terraces with large, open expanses of sand, derived from beach sands and from stream deposition, sand dunes formed on the terrace surface. The climate may have been dryer and colder when these dunes were originally formed. Manchester Beach is a sizable sand dune area, with the source of the sand derived from the Garcia River. The dunes are now fairly well-stabilized by coastal grasses. There are two small areas of dunes on The Sea Ranch: one on the north end, west of Leeward Road to the bluff edge; the other adjacent to Walk-On Beach. There are many other areas along the California coast which have sand dunes. Bodega Head is covered with a long section of dunes which connect with beaches north of town. One of the most spectacular areas of large dunes can be seen near Monterey and Fort Ord.

The older the terrace, the more it has suffered erosion, sometimes to the point of completely removing all traces of it. Because of a lack of fossils, exact ages and correlations from one area to even adjacent areas are sometimes difficult to ascertain. In other parts of the world, terraces have been dated from fossil corals. Sediment cores in the deep oceans and ice cores from Greenland and Antarctica have established good timelines for glacial cold periods and warm periods, and thus help in correlation.

Along the western coast of California, Oregon, and Washington, plate movements of subduction along with the transform movements of the San Andreas Fault and adjacent faults have caused uplift of our coastal areas at different rates. The lower terrace, north of Alder Creek, where the San Andreas Fault goes out to sea, is higher than the same terrace located south of the fault. Uplift rates can be calculated by comparing the current elevation with the age of the terrace formation. These rates seem very small, in the range of .25 to .66mm per year—but over thousands of years, they're significant. On The Sea Ranch, the lower terrace is rising at the rate of approximately .33 mm per year, which calculates as 13" per 1,000 years. The inland area east of the San Andreas Fault zone is also slowly rising. The central portion of the Gualala River watershed, which was the previous site of the Pliocene deposition of the Ohlson Ranch Formation, is rising at the rate of 7.5" per 1000 years.

Terraces on The Sea Ranch

The youngest terrace on our coast is the one currently being formed by wave erosion beneath the slowly-rising ocean. As the waves attack the land, our lower meadow terrace, called Terrace I, is being eroded at an average rate of three to four inches per year.

At the end of the last Pleistocene Ice Age, approximately 15,000 to 18,000 years before present (BP), the sea level was 300' – 350' lower than present. The coastal rivers, during the last glacial period of approximately 100,000 years in length, had downcut their valleys near their mouths some 200' below their current trace. The Gualala River mainstem, from Elk Prairie on the North Fork (site of the Gualala Water Company wells) south some 12 miles to Annapolis Road, was entrenched 160'– 180' below the current surface. Subsequently, with rising sea levels, erosion slowed, sea water occasionally invaded the valley, and deposition of sediments from the ocean and the headwater tributaries has filled the valley mainstem. Other Northern California coastal valleys have similar eroded and filled lower traces up to 200' in depth.

There is also an under ocean wave-cut terrace which has a lower edge at approximately 200' depth offshore, and is located about 1,000' from the current shoreline. This terrace was cut during a warm period approximately 45,000 years ago during the last glacial period. Apparently, there was only partial melting of the continental glaciers, but the melting was not sufficient to raise sea level to its present mark.

It appears the current lower terrace above sea level must have extended offshore for 1,000' or more, and the rise of the ocean has wave-cut into the land that distance. This amount of erosion has occurred in less than 6,000 years. The current rate of bluff erosion averages approximately .3' (e.g. three or four inches) per year. At this rate 1,000' of eroded bluff occurred in 3,000 years. However, the 6,000 year time period since the end of the last glacial period has not been a time of steady warming, but has had at least a couple of cold periods with some regrowth of mountain glaciers, and regrowth of Greenland and Antarctica glaciers. The Little Ice Age in Northern Europe existed for 300 years, ending in the 1850s. With even minor retreat of the ocean level, erosion will slow—so scientists can estimate the cutting of the current terrace has taken 6,000 years. Terrace I averages 1,000' in width, so with continued climate change and sea level rise, erosion could remove most of the terrace in the next 5,000 years!

Terrace I

The lowermost coastal terrace, Terrace I, is thought to have been wave-cut some 80,000–83,000 years BP during a Pleistocene Interglacial warm period. This terrace is the most extensively preserved terrace and varies from zero feet in width (e.g. completely eroded) to 2,000+' (e.g. preserved) in width. The slope of the terrace is gently toward the ocean at 3° – 5°. To correlate from one area to another is sometimes difficult, thus the critical point is the farthest inland point at the head of the terrace. On The Sea Ranch, the highest point on Terrace I is located at approximately 80' – 85' elevation. Sometimes erosion has isolated segments of the terrace into offshore *sea stacks*. There are good examples along the coast, i.e. Goat Rock at the Russian River mouth, south of Jenner, the sea stacks near Elk, and at Gualala Point Island. Each of these rock features have in common flat tops and line-up on the same slope as the adjacent bluff onshore.

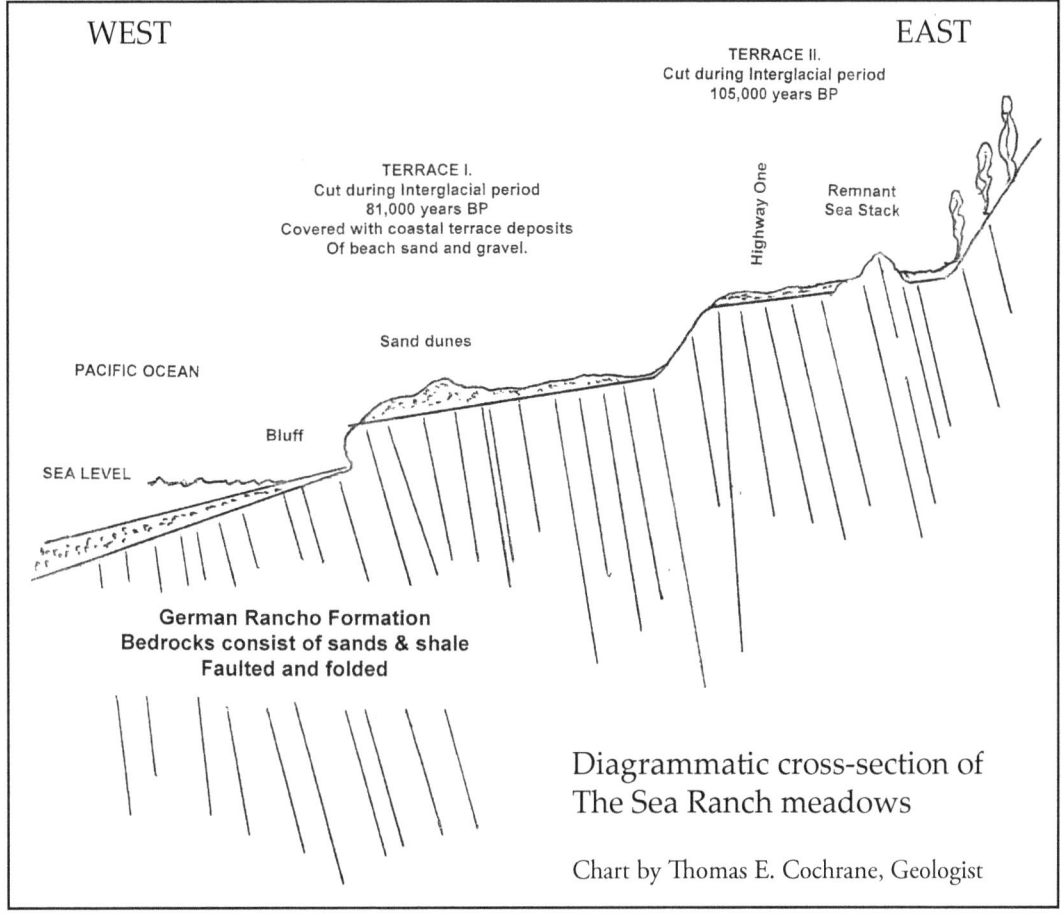

As the sea withdrew from the wave-cut terrace with the advance of a new ice age, beach sands and gravels were deposited on the terrace surface. The result was a smoothing of the surface and a filling in of old streams and depressions. In examining the bluff front at the ocean, one can see the filling of these depressions along with stratification of thin layers of sand and gravel. As is common, the rock terrace is covered by 12' – 15' of sand and gravel. However this may vary from 0'– 35+' in thickness.

Current processes of erosion and deposition of sand and gravel occur each year on our beaches. Winter storms take out the sand beaches and shifting currents in the spring returns sand back. On Walk-On Beach and Clark's Cove Beach, measurements show eight feet of vertical change on average each year. At maximum, 12' of sand is eroded from the beach, leaving just gravel and boulders. With rising sea level during this period of climate change, perhaps less sand may be removed during the winter months. If the storms become more intense, then even greater amounts of sand may be shifted.

Currently, there are many *sea stacks* offshore, but also onshore *fossil sea stacks* protrude above the surface of Terrace I. Some of these stacks have flat tops which line up with the surface of Terrace II. These were hard spots, remnants of Terrace II, left by the ocean when the waves were cutting Terrace I. They occur at several locations: north of Bodega Bay, at Duncan's Landing; south of Stewarts Point; on the south end of Sea Ranch; north of Galleons Reach & west of Sea Gate Road; in Unit 18, west of Mariners Drive; and also east of Sculpture Point Road on The Sea Ranch.

Terrace II

The second terrace east of the ocean is more irregular and more eroded than Terrace I. The date of formation of Terrace II is thought to be 102,000 (102ka) to 105,000 (105ka) years old. On The Sea Ranch, the elevation of the terrace ranges from 100' on the west to over 125' on the east side. Highway One, running through The Sea Ranch, is primarily located on Terrace II. In several spots on the lower terrace, fossil sea stacks protrude above the surface and, in a couple of locations, fossil sea stacks protrude above the Terrace II surface. One such area containing the knobs (fossil sea stacks) is located east of the highway, near The Sea Ranch stables.

The paleo-bluff front of Terrace II, through the northern part of The Sea Ranch, is armored with large sandstone boulders. One can see these along Leeward Road south of Halcyon to Leeward Spur where the rock boulders are found across Highway One. Similar boulders occur in other scattered areas and seem to be related to the armoring of Terrace II when it was the bluff edge with the ocean. There is a good outcrop of similar boulders armoring Terrace II near Timber Cove. The other older terraces do not seem to have similar boulders near their paleo-bluff edge. The boulders remain because they are resistant sandstones. Possibly this marks a still stand (neither rising or falling) of the ocean, and a long period of chemical weathering from the salts of the ocean.

Another unique feature in The Sea Ranch area associated with Terrace II is the presence of a backwater feature on the east edge of the terrace. In some locations, there are slight topographic low areas on the back-slope to the terrace. Carol Prentice (1989) in her thesis near Point Arena, saw a reverse slope of part of Terrace II which she attributed to faulting and folding by the adjacent San Andreas Fault. The Sea Ranch Terrace II back-slope area was formed, *(in my opinion)* by a re-advance of the retreating ocean, which created a lagoon area near the bluff below Terrace III. The lagoon was then partially filled with windblown sand and organic dark soil. Windblown sand is found in the subsoil near Sea Gate Road. After the dune sand was deposited, the area was then covered with beach sand as the ocean rapidly withdrew. The original Sea Ranch water wells were sunk in this area through these water-bearing sands into faulted Black Point Spilite. Another area of marshy lagoon occurs near Long Meadow Road, east of Headlands Close. A similar area also occurs on the east side of Highway One near Timber Cove. At Elk Prairie, on the North Fork of the Gualala River, water well drilling logs indicated sequences of muds and silts interbedded with the sands and gravels, probably lagoonal deposits left after filling the 200' deep channel cut during the previous glacial period.

Terrace III

The third terrace is more eroded, and therefore less continuous than the two lower terraces. Terrace III is found as scattered small benches along the hillsides above the 130'–180' elevation contour. Their age is approximately 133,000 years (133 ka). Most of the original coastal terrace deposits of beach sands and gravels have been removed by erosion, but in many places remnants of sand and gravel can be still found. Houses are clustered on these remnant terraces in that they provide a flatter area to build upon as well as a better view over the steeper slope below.

Terrace IV

The fourth terrace is located with a head slope at approximately 340'–350' elevation.

The age of its formation is approximately 214,000 years BP (214ka). Again, it is more eroded and less continuous. Weathering of the surface has leached the clays and formed a thick soil profile.

Terrace V

The fifth terrace is even more difficult to identify, but occupies spots near the tops of hills, such as near The Sea Ranch Airport. Its age is placed at 320,000 years BP (320ka).

Terrace VI

The sixth terrace is thought to be 390,000 years old (390ka). It is found at approximately 625' elevation. The Gualala Ridge Airport is located on a strip of flat land which is probably situated atop this terrace.

The elevation rise of these terraces calculates an uplift rate of the area ranging from .015" - .019" per year. In her 1989 thesis addressing formations near Point Arena, Carol Prentice stated there were missing terraces in that area having ages of 120ka, 202ka, 305ka and 336ka. The Sea Ranch and other areas in the region have many small, remnant benches scattered at different elevations which may represent some of these missing terraces.

Terrace V and VI have the added complication of being located along the Gualala Ridge trace of the San Andreas Fault. Sag ponds and pressure ridges complicate the picture. Remnant deposits of old beach sand or gravel help us to unravel some of this complexity. Bace Geotechnical has excavated trenches in five spots along the Gualala Ridge Fault on The Sea Ranch, providing us with additional knowledge of the underlying soils and earth materials.

Coastal geological studies by others in the Fort Ross area and further south as far as Santa Barbara, have identified these age terraces and others up to 14 terraces. From a broad perspective, the Pleistocene represents 30 ice ages and 30 interglacial ages in 2.5 million years. As we derive more evidence from glacial studies around the world, the findings become even more complex, especially in light of local anomalies. Even during a maximum period of cold, there may have been a few hundred years or even a couple thousand years of very warm climate. Also, the melting during interglacial times was not complete because the continental glaciers were over a mile thick. However, they responded to small climatic changes, and thus the sea level changed, allowing small terraces to develop.

Rising Land

To address the question posed at the beginning of this section—how did these terraces cut by the ocean end up higher than the elevation they were originally cut at? It seems only logical that the lower terrace was cut when at sea level, and, with no other factors, should have a similar topographic elevation at the present time. But the fact the terraces are now at different elevations indicates the land is also rising. At specific localities, there are several reasons why we find the terraces which were cut at the same time now positioned at different elevations.

The moving Pacific Plate is buckling as it grinds along the North American Plate. As a result, some areas are being depressed into downfolded or faulted blocks, while others are rising. There may even be upwelling convection currents from deep in the earth, pushing the land up. Hot springs and geysers are common inland along some of the adjacent fault zones. The inland areas east of the San Andreas Fault are rising at different rates from the area west of the fault. Since the Pliocene age five million years ago, the coastal ranges have been elevated 2500' or more.

Before the San Andreas Fault became a transform fault, the previous plates were underthrust under the North American Plate. This action is still occurring to the north in Washington and Oregon. This underthrusting resulted in the creation of a string of volcanoes which also mark the eastern edge of the subducted block melting. In Sonoma and Mendocino counties, large outpourings of volcanic lavas occurred, with a notable decrease or lack of violent volcanic eruptions.

The resulting situation is that the preserved terraces, or remnants thereof, while formed at the same time, are now found at different elevations in adjacent locations in this small, 85-mile section of the coast. Various authors studying adjacent areas have arrived at differing elevations for these terraces. In the central part of The Sea Ranch, the five or six terraces exhibit the same amount of uplift—13' per 1000 years. North of Alder Creek on the east side of the San Andreas Fault, the rise of the land appears to be double that of The Sea Ranch area.

Gargoyle formations along the coast, surrounded in the photo above by tafoni.

Special features of interest

Chapter 6

Did God put weird things in our path to confuse us?
OR...to stimulate our thinking?

The following sections and pictures highlight peculiar or unique features of interest found in our area that, at first glance, may not be easily understood. *(I am sure you have your own special spots too.)* There are uniquely shaped rocks along the coast formed by wind and waves. On the north end of The Sea Ranch (as well as other areas) are sandstone outcrops that resemble gargoyles, birds, or seals. *(One of my favorites is one I call "Camel rock.")*

Some of us collect rocks and stones. You also don't have to be a geologist to catch "the bug!" The plate movements have caused quartz veins to be shot through many of the rocks in our region. If you're walking along the beach and see a lot of quartz veining in the rocks, you will discover a fault trace in the vicinity. *(Some quartz veining is irregular in curved patterns, but the ones I like are those in rectangular patterns which show the angles of stress during their formation. With a little imagination, I can see the San Andreas Fault and all the related faults and stress patterns in a rock specimen you can hold in your hand!)*

Different formations erode into various shaped pebbles people like to collect. Some of these have been bored by different types of marine organisms—they make great pencil holders. At low tide, you can see round holes in the rocks with sea urchins living inside. Looking above the tide-influenced zone, you may notice lots of irregular shaped holes, particularly in sandstone rocks. These weathered zones are known as **tafoni**. *(I don't know where this term came from, geologists just like to name things!)* They're formed by chemical and wind erosion—saltwater percolates through the sandstone and softens or removes the cementing material, wind does the rest.

(I have been looking (all my life) for a perfectly spherical rock. There don't seem to be any! Mother Nature produces abundant egg-shaped rocks, but no spherical ones. Maybe you'll be luckier.)

There are few fossils in the area, but some of the rocks show evidence of abundant worm borings which are typically of different color or different shades than the surrounding rocks in which they occur.

BURIED STREAM at Timber Cove

The valley floor here is 15' - 20' wide and flat. The stream has been partially eroded up to the surface. In the stream trace, logs were placed during logging operations in the late 1800s. Some of the logs are situated parallel to the stream trace, and in some areas they're perpendicular. These were used as a corduroy road to transport logs down the valley. The slash and debris from logging was left in the stream valleys, burying the water trace. Now, 200 years later, the debris has decomposed and the stream is beginning to resurface.

Underground streams

Are there spelunking opportunities along the coast?

A stream flowing underneath the surface of the land is a fairly rare occurrence. But in our area there are several underground streams caused by four different processes.

I would venture to guess many people have been in or seen at least one underground stream; I've personally been in an underground stream beneath a glacier in southeast Alaska. There are some spectacular under-glacier streams in Greenland. Both Antarctica and Greenland possess under-glacier streams of fast-moving ice (versus solely water) which flows underneath the main ice sheets. I have also seen lava tubes in Hawaii which once carried molten lava. Possibly you have seen some of those too.

The first underground stream deposit I encountered as a boy in upstate New York was a strange, meandering sand and gravel deposit located in a valley but rising 20' above the valley floor. A road called the Hogsback Road was constructed along the top of the curved surface for nearly a mile. It got me wondering what it was, and how stream deposits could be formed above the valley floor? I became interested in geology, particularly glacial geology, and subsequently learned the feature was an esker, which was formed by a subglacial stream in a crack beneath a glacier. Glaciers once filled all the valleys and covered the land in upstate New York. The Finger Lakes were previous stream valleys deepened as much as a 1000' by the moving continental glacier.

The most common underground streams are found in limestone terrains. Perhaps you've visited a limestone cavern, such as Carlsbad in New Mexico, Mammoth Caverns in Kentucky, or Endless or Lurray Caverns in Virginia. Many states have caverns cut by groundwater and underground streams.

We don't have any of these types along our Sonoma-Mendocino coastline. However, there are more than a dozen partially-underground streams on The Sea Ranch, and several located up the coast in Mendocino County.

Type 1. Underground streams below Terrace I in The Sea Ranch Meadows

In Deep Time, before Terrace I was formed 83,000 years ago (83ka BP), the rising hillside was dissected by streams flowing into the ocean. When the ocean retreated during the last glacial period, the streams which previously just dumped into the ocean poured out onto the newly-formed sand and gravel beaches. If you look at the streams that currently pour onto Walk-On Beach, you will note how the water does not reach the ocean but sinks into the porous sand instead. The same process, on a larger scale, still goes on along Highway One at the upper end of Terrace I. The wet weather streams simply sink into the meadow and flow or seep their way through the coastal terrace deposits which cover the underlying bedrock on their way to the sea. Some streams, including Salal Creek, are strong enough and carry enough water to have eroded the coastal terrace deposits and have created a surface stream which is cut down to the bedrock.

As you wander along the bluff, you will see many of these streams or encounter strong seeps exiting the bluff. Some of these streams have a surface trace near the bluff but have not yet eroded their way to the hillside. Most of these streams which have not cut through the bedrock to the ocean level produce many scenic waterfalls during the wet season.

Type 2. Partially-covered streams by sand dune action

On The Sea Ranch near Walk-On Beach, there is a windblown dune field. A couple of streams occur as surface traces east of the dunes, and then sink beneath the dunes on their way to the ocean. Windblown sand consists of very fine well-rounded grains of sand that are nearly the same size. The porosity and permeability of windblown sand is very high, allowing the water to easily percolate through it.

The major dune area along the Mendocino coast is Manchester Dunes, which blocks and diverts several small coastal streams. Water percolating through the coastal terrace seeps and flows under the dunes in several places. The Garcia River carries enough force to penetrate the dunes, but has been diverted to the south edge of the dune field.

Type 3. Covered streams caused by human activity

In early logging practices, streams were often used as logging "roads." Earth materials, wood, and debris were graded into the streams and often skid roads were constructed to faciliate pulling the logs down off the hillsides.

Surface water and ground water accompanied by debris percolated beneath these log roads, seeking their old channels. If you walk down these stream valleys today, you will note that many of them carry little or no surface water. However, if you listen carefully, you can hear water running beneath the surface—and further downslope near Highway One, the stream may regain a surface trace.

If you take the trail along the Gualala River north of the Hot Spot, you cross several of these streams which are partially covered with earth and debris. In a couple of areas, deep holes extend down to the buried stream, which is more than 10' below the surface. Watch out for these! Near the North Treatment plant along the upper reaches of Salal Creek, redwood logs were placed in the stream forming a corduroy road. These are finally being eroded to the surface.

Similar roads were constructed throughout the area. I have seen one old road at Timber Cove where, in one section of the stream, the logs had been placed across it, and, in an adjacent section, the logs were placed parallel to the stream valley.

Type 4. Underground streams near the Point Arena Lighthouse blowholes

Blowholes develop along cracks or faults in the bedrock beneath thick coastal terrace deposits of loose sand and gravel. Wave action erosion forms a sea cave and, coupled with groundwater percolating along an open crack, a stream is formed within the cave. The stream carries away the overlying loose sand and gravel, and a sink hole develops as the roof gradually falls into the stream.

Bowling Ball Beach

You can only see the bowling balls at low tide!

A unique occurrence of round, sandstone boulders is found at Bowling Ball Beach. Parking is available on the west side of Highway One, just north of the Schooner Gulch Bridge at MEN 11.5. The trail to the north accesses the beach down a difficult rope ladder located at the base of the bluff trail. (*The basal section of the ladder has been recently replaced with a device designed to shift when battered by winter storms. Many of the beach access stairways along the coast are partially destroyed each winter by the violent storms. We hope this fix works.*) One may also access the beach a mile north at MEN 12.40 near Ross Creek. There is a parking area on the east side of the Highway, near an emergency telephone stop. A small obscured trail with easy access to the ocean opens just north of the parking area.

Along this section of the coast, the rock formation in the bluff front dips steeply west at approximately 65º. The formation is named the Galloway-Skooner Gulch Formation of Miocene age. Note this early spelling of Schooner, referenced on old maps as Skooner Gulch. The formation forms a dramatic bluff cutting diagonally across the beach area, exposing bed after bed. The harder layers are sandstones and siltstones interbedded with grey shale. All the beds are quite thin. Erosion is cutting out the thin, soft shale beds and the harder beds are then more pronounced.

The bowling balls occur in a short segment of beach area just north of Galloway Creek, immediately north of Schooner Gulch. The bowling balls can only be seen or accessed at low tide. They are round sandstone boulders approximately four feet in diameter in five or six rows, with 15 or more per row. The underlying rocks are shale and siltstones which steeply dip westward.

The boulders line up in troughs between the harder ridges.

So, the big questions are these: 1.) how were they formed? 2.) why are they close to each other and in straight lines?, and 3.) are they eroding out of the bedrock formation, or have they been rolled around and deposited here by the ocean?

This group of boulders begins at a fault on the south, the Galloway Creek Fault, and ends near another fault to the north. In the mile section south from Ross Creek to Galloway Creek, there appear to be only four faults, and these are marked by notches in the bluff. However, all of the rock strata are crisscrossed by two or three sets of joint patterns. Remember, joints, in contrast to faults, are cracks with little or no movement and are caused by expansion of the rocks as pressure is released from erosion of the overlying confining rocks. They may also be caused by pressure of the moving plates along the San Andreas Fault. The softer shale breaks up into very small fragments and is highly susceptible to landsliding.

Some people believe the bowling balls were rolled around by the ocean and left in this particular spot. Then why are the rows so regular and the size of

"Hamburger buns" at Bowling Ball Beach

the balls so similar? Some of the balls are attached to the underlying rocks and even appear to be weathering out of a particular layer. But could they be bound in place by natural cementing onto the underlying rocks? In many instances, round boulders are concretions of minerals precipitated, layer by layer, around a hard core, often a fossil. However these large bowling beach boulders show little evidence of layering, and we don't have one that is split to the center to study. So this is a puzzle to put to your geologist friends.

As we proceed northward along the beach, there are more strange rocks which are definitely eroding out of the rock formations. Several of these look like giant hamburger buns! They are yellow in color from iron staining. Their size varies from 4' – 10' in diameter and they are up to 4' thick. Further to the north are smaller, spherical balls 1' – 2' in diameter. These rocks overlie layers of shale and siltstone which have been highly-burrowed by marine worms. The boulders themselves also contain worm tubes. Some resemble branch-like tree limbs, but are probably intersecting worm tubes. Some areas have darker sandy zones which are also bored by marine worms. These questions remain: were these nests of worms, or algal patches that later were bored by worms?

On a smaller scale, around the corner to the south near Schooner Gulch, there are thin sandstone layers that are jointed in one to two inch cubes. Wind and water have weathered these cubes into spherical shapes which are still attached to the main rock strata. Is it possible the bowling balls are a larger example of this type of weathering at a location where the joints are further apart?

When you visit the area, take a look at all of the three or four types of "balls" which are present, and give us your opinion.

Devil's Punchbowl—related sea caves & holes near the Point Arena Lighouse, Mendocino County, CA

Do we have to worry about sinkholes under our houses?

How can the rather unique set of "holes" in the bluff near the Point Arena Lighthouse be explained? These holes are up to 50' in diameter and up to 50' deep. These are located within a few feet of the bluff edge, but appear to be developing as far as 150' from the bluff edge.

The underlying bedrock in the area consists of a series of thin-bedded sandstones, siltstones, and shale belonging to the Point Arena Member of the Monterey Formation of Miocene Age. They were deposited approximately 20 million years before present (20ma BP). These rocks have been folded, faulted, and uplifted above the ocean by movement of the Pacific Plate along the San Andreas Fault. Since deposition in the Santa Barbara area in a deep ocean basin, this section of the plate has moved approximately 300 miles northward to its present location.

In the area near the lighthouse, the rocks dip southwest from 2º to over 56º. At Arena Cove, the rocks are further twisted into a prominent anticline. The result has produced highly-fractured rocks with prominent joint patterns in two or more directions. More intensely fractured rocks are easily eroded by groundwater and ocean waves. *(See the Road Log Appendix for folding and faulting at Moat Creek which has the most intensely squeezed rocks in the area.)*

The land surface has been further modified by coastal uplift, sea level changes, erosion, and deposition of younger sediments. During Glacial times, and especially in the past 600,000 years since the last major Interglacial period (the Sangamon), five or more coastal terraces have been formed by changes in sea level. Each of these terraces indicates a long period of wave erosion and encroachment of the ocean onto land. Sea levels rise during interglacial times and fall during glacial times.

As the seas retreated, beach deposits of sand and gravel were left covering eroded bedrocks. The lowest terrace and closest to the ocean, Terrace I, is thought to have been formed 83,000 years before the present (83 Ka BP) at a high stand of the ocean during an Interglacial period. The bedrock surface was eroded by ocean waves to a relatively flat surface with a gentle dip toward the ocean. Streams from the adjacent lands dumped sediment into the ocean. Oscillations in sea level allowed major streams to cut valleys into the bedrock.

As the seas withdrew for the last continental glaciation (50 to 25 Ka BP), beach sands and gravels filled in the valleys and covered the bedrock, leaving a smooth surface. Sea levels remained low until about 10,000 years ago (10 Ka BP) when the glaciers melted and the ocean rose to the current sea level about 6,000 years ago.

Ocean waves beat against the coastline and erode the least-resistant rocks, forming the current headlands, coves, and islands. Numerous sea caves were carved out, especially in the Point Arena area. The sea caves commonly occur along small adjustment faults, or in intensely fractured areas. The holes in the bluff—such as the **Devil's Punchbowl** at the Point Arena Lighthouse, and **Satan's Hole** and the **Witches Cauldron Hole** located a half mile to the south—are the result of roof collapses of the developing sea caves.

The sea caves allow waves to erode and penetrate into weak areas with thick coverings of sand and gravel left as coastal terrace deposits. The resulting debris from the erosion is carried out of the caves by retreating waves. Once the cave has eroded the ceiling bedrock, exposing the overlying weak coastal terrace deposits, gravity causes the sand and gravel to fall into the caves. Ocean waves quickly remove this material. It's interesting to note that the sinkhole might be quite large, 50' in diameter, but the cave opens to the ocean by way of a small hole 10' or so in diameter at the base of the bluff. Ultimately, the holes will be connected to the ocean as the roofs fall. Sea stacks, coves, or islands will be the result. Current and historic photographs of the Devil's Punchbowl at the lighthouse show this dramatic change from a hole to a dissected point of land.

The following maps, diagrams, and photographs illustrate the above process for the development of these "holes."

A series of large sink holes are found south of the Point Arena Lighthouse. Short streams which developed along faults caused sea caves to form. Groundwater percolated down through the porous overlying sediment, dropping material into the caves. The waves subsequently removed the collapsing sand and sink holes resulted.

78 Thomas E. Cochrane

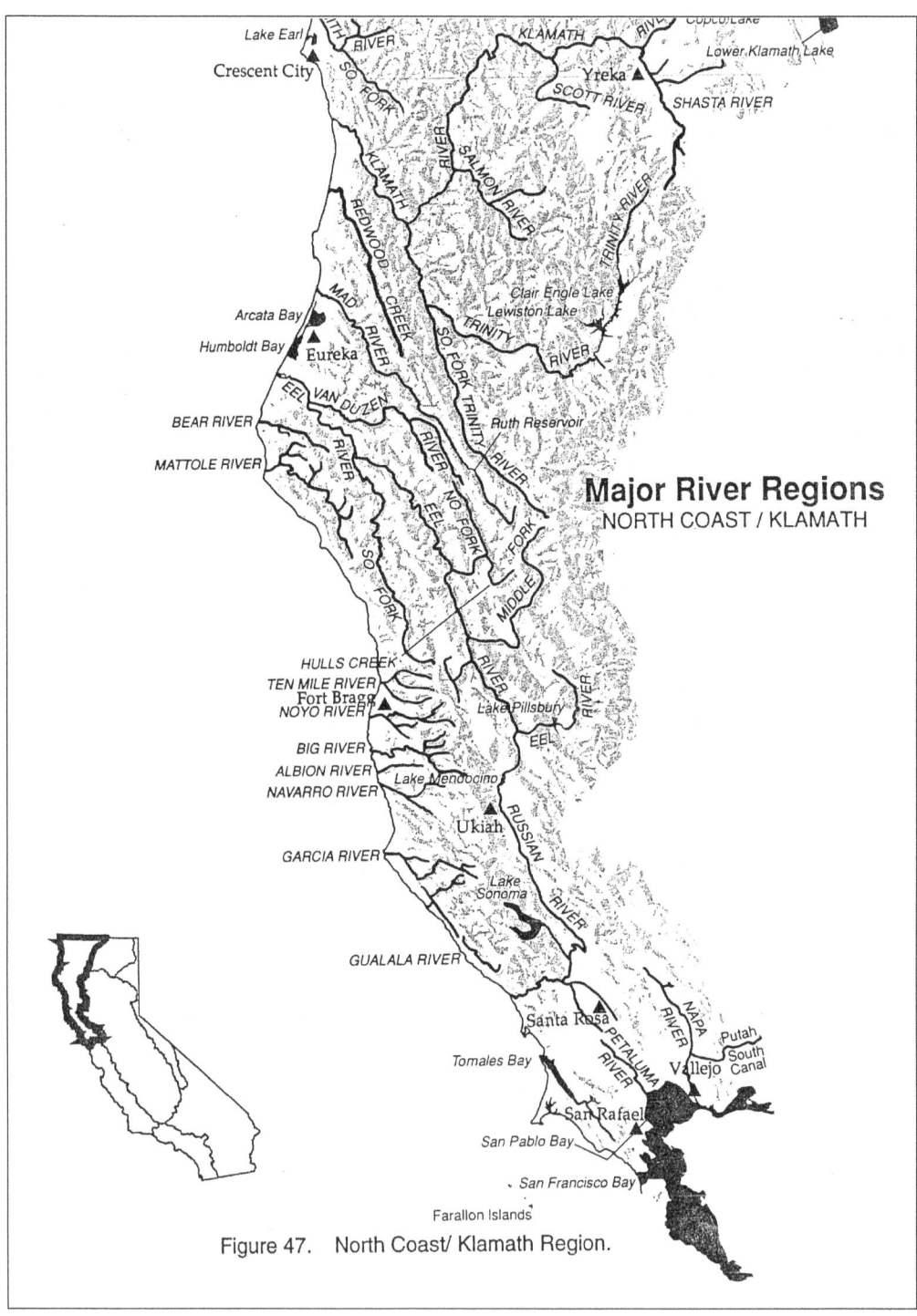

Figure 47. North Coast/ Klamath Region.

Source: California State Lands Commission, 1993

Coastal River Watersheds:
Eel, Navarro, Garcia, Gualala, and Russian Rivers

Chapter 7

What are the threats to these watersheds and how can we protect them?

Collectively, the above five rivers drain nearly 6,000 square miles of watershed along the Sonoma, Mendocino, and Humboldt County coastlines. Sediments carried by these rivers, and their ancestral rivers, have been deposited offshore into subsiding basins which have existed since the Cretaceous Period. Historically, these rivers have been the habitat of salmon and steelhead. Sadly, fish populations have declined drastically due to warming of the ocean and stream waters, building of dams, and from excessive sediment discharge. Clearcut logging practices have warmed the land, produced excessive erosion, and removed tree canopies shading streams. More recent logging practices have slowed or reversed some of these environmental effects. However many of the streams are still choked with sediment, affecting the number and depths of stream pools that are necessary to raise and sustain the fisheries. The current trend is to replace the timber forests with vineyards, at least in many of the headwaters. This cuts down the forest canopy, raises the land temperature, increases sediment runoff, and uses large volumes of water.

Coastal rivers have adjusted their courses to the topography, which is controlled by folding and faulting of the coastal ranges. As the land slowly uplifted from the ocean, surface waters ran directly into the ocean at near right angles to the land. The streams which result from the rising land are called *consequent streams*. Later, some of these streams cut through the closest coastal range and captured streams flowing behind that range. The Gualala River lower section, including the estuary, is one of these. It is thought that the Gualala River and the Garcia River may once have drained as a single stream to the northwest, and dumped into the ocean near Point Arena. Subsequent uplift of the watershed and landsliding raised and filled a section of the mainstem near the Fish Rock Transfer Station as much as 350'. The effect was to split the old watershed into two separate watersheds.

If you examine stream patterns in these watersheds, you will note that many of the tributaries have a very rectilinear pattern following the various blocks of the coastal range. This is known as a *trellis stream* pattern. Some areas have a more *dendritic* pattern, like the veins in a leaf.

Many area valleys have very steep slopes which are subject to landsliding. The steepness suggests earlier rapid erosion and downcutting of the land. These processes point to rapid uplift with heavier precipitation, compared to what we are experiencing in the current climatic period.

Eel River

The Eel River's mouth is in Humboldt County, 15 miles south of Eureka. Since this study does not deal with the coast that far north, you might ask why is it being included here? The Eel River watershed is the largest watershed in northern California, encompassing 3,684 square miles. The headwaters of the Eel are a few miles northeast of Ukiah in Mendocino County. PG&E utilizes a diversion of waters for electric power from Eel River tributaries at Potter Valley, which directs Eel River water into the Russian River system. (This historic diversion was dug by Chinese laborers.) Environmental groups, including Native Americans, have tried to stop this diversion for many years. These combined efforts have accomplished the removal of some dams on the Eel River, and also changed some of the salmon fishing rules that apply to Native Americans. The Sonoma County Water Agency and some Russian River environmental groups don't want the diversion stopped as it provides water for Santa Rosa and helps maintain flows in the Russian River which are beneficial to the fishery.

Lake Pillsbury collects water in the upstream headwaters which then flows into Lake Arnsdale, then flowing through a tunnel from the power plant into Potter Valley, which then flows into Lake Mendocino at Ukiah, and finally drains into the Russian River. All three lakes are manmade and controlled by dams.

Navarro River

The Navarro River watershed drains 315 square miles into the Pacific Ocean, eight miles south of the town of Mendocino. The lower watershed east of Highway One, and along Highway 128 flows through a preserved redwood forest, Navarro State Park, which is a marvelous drive on the way to the wineries of Anderson Valley.

Garcia River

The Garcia River watershed drains 144 square miles, with a long mainstem that follows the San Andreas Fault. The mouth is located just north of the Point Arena Lighthouse today, but was once probably further north near Alder Creek, where the San Andreas Fault goes offshore.

The lowland around the mouth of the Garcia is a broad plain, possibly down faulted, between the Hathaway Creek Fault and the San Andreas Fault. It has long been owned by the Stornetta family and recently made a conservation preservation area. Seasonally, Snowy (white) Egrets land on the flat near the highway during their migration south.

Gualala River

The Gualala River watershed drains 298 square miles east of Gualala and The Sea Ranch. The watershed is 30 miles long and up to 20 miles wide. The boundary between Sonoma and Mendocino counties is delineated from the mouth of the river, along the estuary, upstream to the Green Bridge. The mainstems of the North Fork and the South Fork follow the San Andreas Fault zone, and line up with the mainstem of the Garcia River located to the north.

It is possible the two rivers' watersheds were joined as a single watershed, and later separated by uplift occurring near the current Fish Rock Transfer Site? Also, the Gualala River estuary section is steeply downcut through the first coastal ridge—it may have been superimposed on the landscape of the moving and rising Gualala Block in the distant past. At the rate of movement of the Pacific Plate along the San Andreas Fault, the current mouth of the Gualala River would line up with where the Russian River is now around 3.8 million years ago.

Russian River

The Russian River watershed snakes south of Ukiah, through Healdsburg, to the Santa Rosa Plain, then west through Guerneville to the mouth at Jenner. The watershed covers 1485 square miles.

Most of the coastal rivers in northern California are located primarily in rural areas with a low population density and little agriculture, other than tree production and growing of wine grapes.

The Russian River is heavily populated with a string of communities from Ukiah to

Jenner. Much of the adjoining land is currently in vineyard production. Previously, prunes, apples, and hops were grown there. The watershed is a major area of tourism. Gravel mining in the mid-reaches of the river has been extensive. Water wells dot the area, serving vineyards and mini-ranches. Water usage is high and dams hold water in Lake Sonoma, and in the previously mentioned lakes capturing water from the Eel River. Water is sold outside of the watershed to cities

Watershed of the Gualala River. Map courtesy of the Gualala River Watershed Council.

south of Santa Rosa. Waste water from Santa Rosa is currently piped to the Geysers Geothermal Field and used for steam generation of electricity.

The Warm Springs Fish Hatchery has been greatly impacted by all this human activity, and the numbers of salmon are very small. Environmentalists monitor the river and actively battle further development. This fish hatchery was constructed as a mitigation for the building of the dam and formation of Lake Sonoma and has helped to stock fish in the river beyond the natural native fish runs.

Gualala River watershed resources

Can humans use this water, or should we save it for the fish?

The Gualala River watershed of nearly 300 square miles is situated adjacent to the coast, west of the highly-developed Russian River corridor of vineyards, local residents, and commerce. It is sparsely populated by farmers, loggers, and environmentalists who all resist further development of any kind within the watershed. A few areas of wine grapes have been planted in recent years. Friends of the Gualala River and other organizations have recently defeated the application for development of Preservation Ranch into mini-ranches and vineyards near the Annapolis community.

This area was chosen by the State as the first in Northern California for a watershed assessment program. A Gualala Watershed Council was formed consisting of owners and environmentalists interested in the program. Various restoration activities are ongoing, aimed at protecting the fishery and re-establishing many of the tributaries for steelhead habitat.

Much of the area near the coast is actively harvested for Redwood and Douglas fir trees. Some inland areas have been logged in years past and converted to cattle-grazing ranches. Recently, some of these areas have been converted to grapes. Gualala and The Sea Ranch are adjacent to the estuary and mouth of the watershed. An estimated 4,000 homes along the coast use water from the North Gualala Water Company and The Sea Ranch Water Company. These water companies have restricted water diversions during low-flow periods. Sea Ranch has a reservoir with a two year capacity of water storage.

Approximately one-quarter of the watershed rests in Mendocino County and three-quarters in Sonoma County. At the coast, the estuary from the Green Bridge to the ocean forms the boundary between the two counties.

The San Andreas Fault Zone is located at the western boundary of the watershed, behind the first coastal ridge. Ten and one-half miles of the mainstem located along the fault is at less than 56' elevation. The floodplain in the mainstem is up to 1000' in width. Previous erosion during glacial times along the mainstem cut the valley floor as much as 200' below its current level. Subsequent rise in sea level and erosion in the tributaries have partially filled in the valley with thick layers of sand, gravel, and mud.

Five major tributaries have cut through the coastal ranges and eroded the watershed from elevations in the headwaters of 2,000'–2,600'. The sediment carried by these streams dumps into the mainstem low gradient section of the river. Sediment transfer to the ocean mainly occurs during significant rainfall events when flows in the lower reaches elevate the water level by 5'–10' above dry season levels.

Estimated water from runoff in the watershed

The 300 square mile Gualala River watershed receives an average estimated 36" of rainfall per year during the six month rainy period, October through March. Low rainfall years range from a low of 30" to a high of 80+". To calculate the amount of water that falls on the watershed, we take 300 square miles times 640 acres per square mile which equals 192,000 acres of watershed. At 36" of precipitation per year, this calculates as 567,000 acre feet per year.

A high percentage of this rainfall occurs as runoff. Initially each year, the soil is first recharged with groundwater. Later, rain runoff has varying rates of discharge, depending upon the saturation of the soil, and the intensity of the rainfall. Possibly 50% of the rainfall runs off into the ocean during the winter months. The actual percentage of rainfall runoff from the Gualala River might be higher than an average river because of two factors: 1.) the main tributaries are cut deeply into the watershed, and 2.) the primary bedrock formation covering most of the watershed is the Franciscan Formation which has a low permeability and may not absorb water as easily as more porous sandstones and sandy loamy soils.

If we calculate that 50% of the annual precipitation is runoff, this means that 283,500 acre feet of water is runoff from the watershed. However, I note that the Department of Water Resources estimates the annual runoff of the Gualala River is 558,000 acre feet, which is twice as high as our calculated estimate. This number appears to be high, but may represent runoff during wet years.

If we recalculate runoff using 48" (four feet) of precipitation as the average annual rainfall, we get 768,000 acre ac.ft./yr. and with 75% as runoff, it calculates as 576,000 acre feet of runoff per year.

The USGS Gage at Annapolis 1951–1971 gauged runoff in the Wheatfield Fork from 138,031 ac.ft./yr. to 560,214 ac.ft./yr. for an average of 310,400 ac.ft./yr. We would need to add the South Fork and the North Fork figures to these numbers to get a total for the watershed. Obviously more data is required to achieve accurate numbers.

Groundwater in the watershed

Groundwater is trapped within the soils, sediments, and bedrock. The sands and gravels along the mainstem of the Gualala River have high porosity and permeability, and trap lots of groundwater as a result. Porosity is estimated to average up to 30%. The Ohlson Ranch Formation that caps many of the hills also has high porosity, which may average 25%. The Franciscan Formation bedrock (which underlies most of the watershed) varies greatly in porosity and permeability, but probably averages only 10%–15% porosity. This does not count fractures which increase permeability greatly and, to some extent, add to the pore space available for water. Water wells with good productivity depend on fractures and probably have been drilled into fault zones.

Water table is defined as the depth where water is encountered below the land surface. The water table rises as groundwater is recharged from precipitation into groundwater reservoirs. Some amount of precipitation is lost due to evaporation. Groundwater slowly seeps into the streams throughout the year, and is the source of water in the streams during the dry season. Occasionally, groundwater exits the water table in springs, and many of these are scattered throughout the landscape. Commonly, springs occur at the contact (intersection) of two different formations where porous rock encounters more dense rocks.

Vegetation draws water primarily from the upper 15' of the soil. Reportedly, evaporation and transpiration of forested areas in Northern California require three acre feet per year of water. This number seems compatible with what we see in the area. Redwood forests require more than three acre feet of water per year. This limits the range of the redwoods to the first coastal ridge and most of the western part of the watershed. Inland areas, such as Dry Creek and the Santa Rosa Plain, get less than 36" of precipitation per year and have few redwoods.

The majority of water usage by humans in the watershed comes from groundwater. There are some ponds, particularly those constructed recently for vineyard irrigation. These ponds capture and store surface runoff during the wet season for dry season usage by vineyards and cattle. Most residences have a water well withdrawing water from the water table. The North Gualala Water Company and The Sea Ranch Water Company withdraw water from wells within the Gualala River Mainstem.

The Division of Water Resources, Bulletin 118, has a section dealing with groundwater in the watershed. It lists the Ohlson Ranch Formation and the Franciscan Formation but makes little reference to the Gualala River Mainstem, which is the primary reservoir of water in the watershed.

Once water percolates into the ground, only a portion of the water is easily recoverable. Water coats the grains of sand and only the water between the grains is moveable water. We use a term called "Specific Yield" to measure the amount of water that can flow from a formation. For sands and gravels, a common Specific Yield is 20%–25% of the water in place. In 1989, I wrote an unpublished study of the Gualala River Mainstem and estimated the groundwater in place. The mainstem is up to 1000' in width and 180' in thickness, and filled with sand and gravel. The mainstem of the North Fork and the South Fork is 10.5 miles in length. The five tributaries have small areas, up to 500' in width and 50'– 60' in thickness near their junctions with the mainstem. Calculating the volume of this groundwater source, we get 150,000 acre feet of water with a Specific Yield of 25% which calculates a yield of 37,500 acre feet of recoverable water.

The Sea Ranch Water Company has a reservoir with 300 acre feet of storage. The Sea Ranch uses approximately 400 acre feet of water annually. No water is withdrawn from The Sea Ranch wells during low flow periods to protect from drawdown of the surface stream. It is highly questionable how much drawdown occurs at the surface of wells which are drawing water 160' below the surface.

River trace of the mainstem

The mainstem is accessible at the mouth at Gualala, at the Green Bridge which is the junction of the North Fork and the South Fork, at The Sea Ranch Hot Spot, and near Valley Crossing on Annapolis Road. All of these spots on the mainstem are situated at less than 50' elevation.

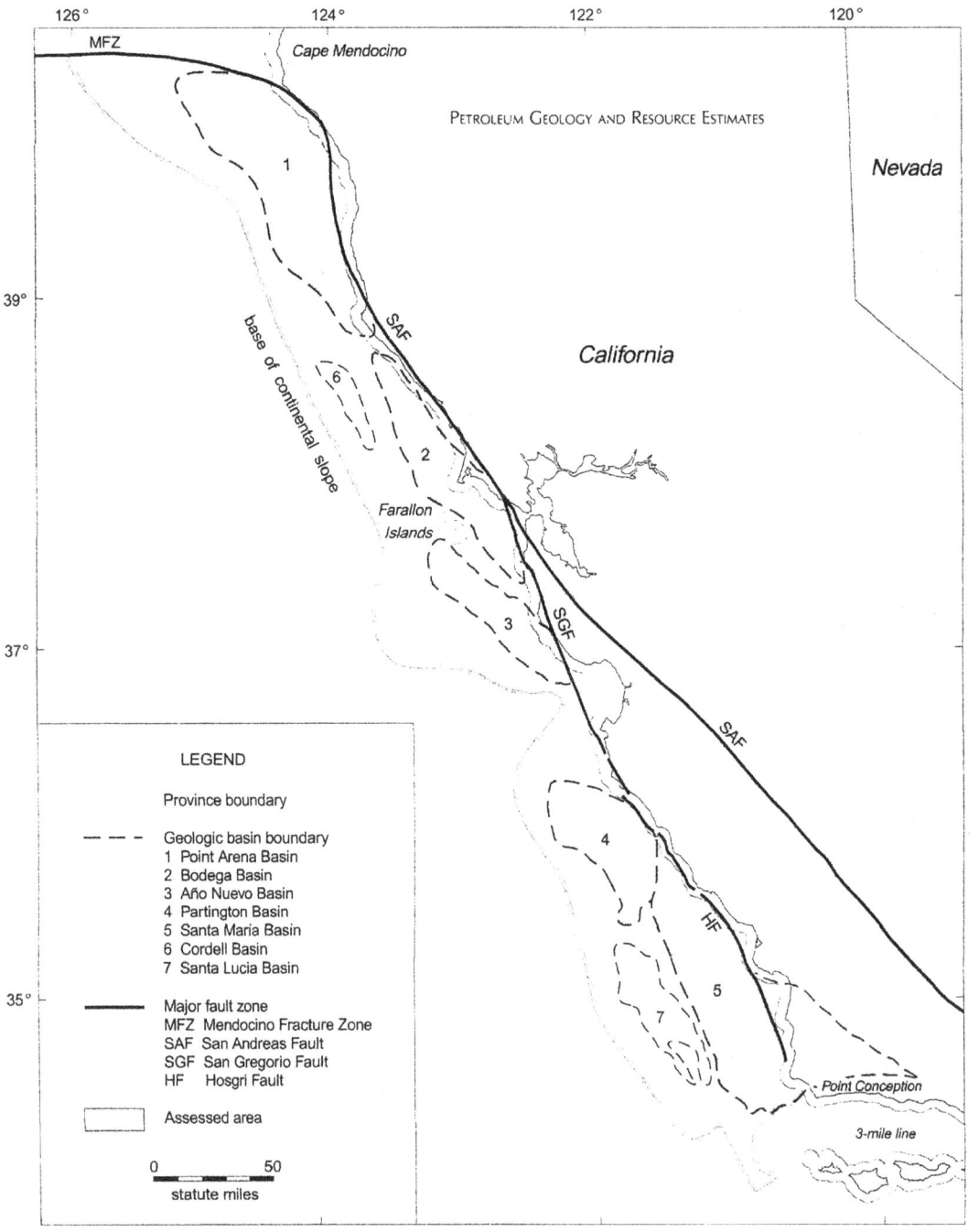

Source: U.S. Dept. of Interior, Map of Central California province showing geologic basins, 1995

Offshore basins

Chapter 8

You mean there's oil out there?

A series of offshore basins exists along the California coastline. From the northern edge of the Santa Barbara Basin area (35° N) to the California Oregon border (42° N), over 480 miles, there are eight underwater basins. At the north end is the Eel River Basin, followed by the Point Arena Basin, the Bodega Basin, the Santa Cruz Basin, the Sur Basin, and the Santa Maria Basin. Some authors use the term the Gualala Basin for the area that covers the northern part of the Bodega Basin. These basins are 20–60 miles wide from the coastal edge, and extend to the lower edge of the continental slope.

We know about these basins because some parts of them are exposed onshore where we can examine the sediments, but large sections are offshore. The offshore areas have been partially explored by seismic surveys and drilling by major oil companies. This wedge of land west of the San Andreas Fault (onshore and offshore) is called the Gualala Block, and of course includes the offshore basins. Earth's crust is thin under the deep ocean, approximately 10 miles in thickness, and composed of basaltic-type heavy, dense rocks. The continents are thicker, up to 30 miles in thickness and composed of lighter rocks. Much of the continental mass is composed of sedimentary rock. The drifting continents ride over the more dense oceanic crust.

The edge of the continent is marked by the *shore*, followed offshore by a shallow area of a terrace being wave-cut or recently wave-cut. Following this is the *continental shelf*, sloping gently to 1,000' or more in depth, where the offshore basins are collecting sediment from the land. At the edge of the shelf is the *continental slope* which plunges down to nearly two miles in depth to the deep ocean. At the base of the continental slope is the primary zone of *subduction*. Most of the subduction in this area (Sonoma and Mendocino counties) has been translated into transform motion along the San Andreas Fault. However, along the Oregon and Washington coast, subduction is still actively occurring.

Various energy companies have eyed these offshore basins for many years, knowing their potential for major oil and gas extraction, and several exploratory offshore wells have been drilled with what they consider to be promising results. Tar balls are occasionally found on our beaches.

Most of these are probably from passing ships, but some may have come from natural seeps. A point of history: when Sir Francis Drake explored the Santa Barbara Channel area, he noted in his log book that black oil was floating on the surface. The Santa Barbara Basin area is broader and possibly more complexly faulted than the basins to the north, although we don't have the well record evidence and seismic evidence to complete the picture.

The Bodega Basin and the Point Arena Basin extend along the coast offshore from Half Moon Bay on the south to Cape Mendocino on the north. Shell Oil has drilled 11 offshore exploratory wells in these basins. The Bodega Basin is approximately 110 miles long from Half Moon Bay to Gualala, and 20 miles wide offshore, with an area of approximately 1,700 square miles. (As mentioned, some writers refer to the Gualala Basin. We could split the Bodega Basin at Black Point, which is an anticline, then Bodega Basin would be located to the south, and the Gualala Basin would be north of Black Point.)

Estimated thickness of sedimentary rocks is in the range of 20,000' above basement rocks. The Point Arena Basin extends 100 miles from Point Arena to Punta Gorda and is 30 miles wide offshore, with an approximate area of 3,000 square miles. Estimated sedimentary rock thickness is in excess of 15,000'.

The rock formations in the basins are similar in age and lithology to formations producing oil and gas in southern California. Potential oil and gas reserve estimates vary from many billions of barrels to trillions of barrels of oil, with like amounts of natural gas. With only 11 exploratory wells in an area of 4,700 square miles, an accurate assessment of reserves is impossible. Exploration was undertaken in the late 1970s after the OPEC oil embargo. Assessments in the late 1990s indicated exploration was warranted with a price of $50 per barrel of oil at that time. When oil prices climbed up to $100 per barrel in recent years, more exploration for oil and gas was warranted. Discoveries and production in North Dakota have filled the need for increased domestic production and the price has since plummeted. Californians don't want offshore oil platforms for fear of oil spillage, which is rare but disastrous. *(Personally, I worry more about offshore oil tankers traversing our coastline. Shipwrecks have been common along our coast and have been a greater source of pollution worldwide than from oil platforms.)*

Seismic data indicates that a series of offshore faults run parallel to the coastline and the San Andreas Fault. Periodically, earthquake shaking is felt from minor movements on these faults. The onshore rock formations in the Gualala Block are buried in the bottom of the offshore basins. However, as you might expect, the uppermost rocks and sediments have been derived from erosion of the land. Several thousand feet of Pleistocene and Holocene sediments have been measured in the offshore exploratory wells. The Garcia River and Gualala River watersheds have been uplifted over 2,000' since the Pliocene Ohlson Ranch Formation was deposited less than five million years ago, and are the sources of the sediment.

Submarine canyons

How do canyons develop under the ocean?

Submarine Canyons look like river channels, but are found under the ocean on the continental shelf and the continental slope. They have developed in many places off the California coastline. The most well-known one is the submarine canyon located just offshore from Monterey Bay which is also the most well-documented.

Beginning north of Cape Mendocino and proceeding south, we have the Eel Canyon, the Mendocino Channel, the Viscaino Canyon, the Noyo Canyon, the Navarro Canyon, the Arena Canyon, the Bodega Channel, the Cordell Channel, the Farallon Channel, the Pioneer Channel, and then the Monterey Canyon.

These undersea submarine channels are cut into Pleistocene and Recent sediments. They developed during the Pleistocene, or possibly even earlier, during periods of low sea levels. These were glacial times of high precipitation and increased sediment loads. Onshore streams poured onto and over the continental shelf, bringing large loads of sediment. Longshore currents piled up and collected sediments at the edge of the shelf. Storms or earthquakes caused submarine landslides and *turbidity currents*. (A turbidity current is a current of water carrying large amounts of silt, sand, and gravel. Because of the sediment load, the current flows downslope under the ocean.) Once a canyon is initiated from one of these flows, subsequent undersea slides and turbidity currents keep the canyons clean and produce more erosion, thus deepening the channels. These processes are ongoing and have been documented at many places around the world. Turbidity currents have cut many undersea communication cables and registered speeds up to 100 mph.

These canyons are up to 3,000' deep, a half mile wide, and several miles in length. Some of these canyons show right angle bends at several locations. The apparent offsets are caused by faults having strike slip movement. Within the offshore basins, there are strike slip faults which run parallel to the San Andreas Fault. In the Gualala Basin, the Black Point Spilite has a strong magnetic character. Offshore of Del Mar Point on The Sea Ranch is a magnetic high that appears to correlate to the magnetic high over the Spilite at Black Point. If these two magnetic features are related, the offshore fault has a mile or more of offset.

The Noyo Canyon is of particular interest because its head abuts the offshore trace of the San Andreas Fault. If we slide the Noyo submarine canyon south along the San Andreas Fault at one inch per year, then the Noyo would have been opposite Point Arena 2.5 million years ago—continuing the restoration of the slip south to the Gualala River mouth would take another million years. Further slippage to the current Wheatfield Fork position is another half million years to a sum of four million years, which is the Pliocene Age. You will recall that during the Pliocene, the inland area, now the Gualala River watershed, was a marine embayment where the Ohlson Ranch Formation was deposited. We don't know whether the Noyo submarine canyon existed at that time or developed later during the Pleistocene. The beginning of the Pleistocene would have marked a drop in sea level and the drainage of the Ohlson Ranch Pliocene embayment. An outflow of sediment from the draining basin would have possibly generated many turbidity currents in the area, one which could have started the development of the Noyo Canyon.

If these canyons date back several million years as suggested by the Noyo Canyon, then considerable sediment has been carried off the continental slope into the deep ocean. This may mark the development of new offshore basins which will be subducted beneath the continent in the future.

Shaping the Sonoma-Mendocino Coast   93

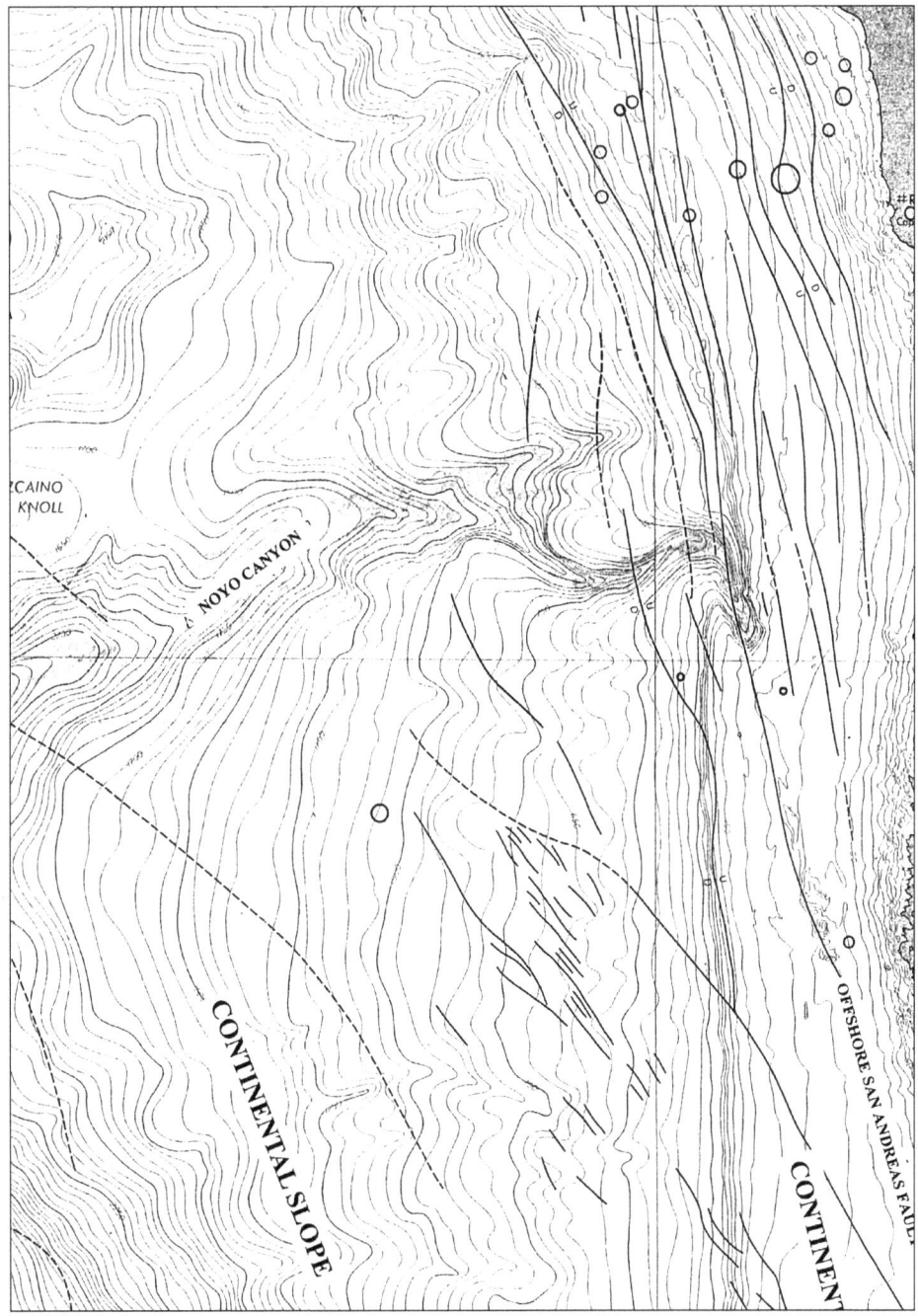

Noyo Canyon lies eight miles offshore from Fort Bragg. The San Andreas Fault causes the canyon to turn four miles south along its trace. The Continental Shelf in this area is 8-10 miles in width. The Continental Slope descends over 18 miles to the deep ocean at more than 10,000' depth. Source: the map is a modified portion of the CA Division of Mines & Geology, Map 5B of the CA Continental Margin Geologic Map Series, 1989.

This is an old photo of early logging which took place from the 1880s to the commencement of tractor logging in the 1920s. Much of the area was clearcut and the creek beds were turned into logging roads. The buried slash and corduroy logging roads are finally decaying to the point where the streams are returning to surface traces. Much of the area was burned after the logging. The advent of tractor logging allowed more roads on slopes, and thus the streams were better protected. However more logging roads were cut, producing sediment sources and further eroding the environment. The mainstems of our coastal rivers are all choked with sediment.

Modern Forest Practice Rules were implemented to protect the environment from further degradation. However, clear cutting is still the current practice of many of the logging companies. These rivers still run brown after each rainfall.

Human impacts on our environment

Chapter 9

We have changed the face of Earth and are now living with the consequences.

In the last 10,000 years, humans have altered our planet to suit our own needs. We have tapped the world's resources for energy and burned up a large percentage of the coal, oil, and natural gas which were formed millions of years ago in long, slow, natural processes. We have cut many of the world's forests to obtain wood for buildings, furnishings, and heating. We have developed agriculture, replanting the natural environment with clones we have created. We have harvested the natural environment and fished the oceans.

Humans have terraced the landscape, dammed the rivers, constructed levees along many of the major rivers, built jetties into the oceans, dredged the rivers and harbors, and changed the courses of rivers and ocean currents. Cities and their infrastructure are expanding over the land. Pipelines carry oil and water from one area to another facilitating new population centers.

Our factories, power plants, and automobiles pour pollution into our atmosphere, rivers, and oceans. We have sent ozone into the stratosphere and Earth is now encircled by satellites and space junk. Nuclear waste we've generated has made some areas uninhabitable for hundreds of years.

We have wiped out many species of plants and animals. We spray chemicals everywhere to wipe out unwanted bugs and vermin. Currently, nature is experiencing a long dying period. The coral reefs, the rainforests, the deltas of the major rivers are all threatened. Invasive species have been brought by ships to ports throughout the world and are destroying native ecosystems.

This book has primarily discussed the natural geological environment along this specific coastal region which, at first glance, is only minimally affected by the above-mentioned issues. However, we have logged much of the area's forests, planted cultivated species, built roads and houses, cleared forests and brush, over-fished our ocean, and misused some of the natural landscape here too. And we have learned that what happens on the other side of the world does affect us locally as well.

Humans certainly are a major destructive force on the environment, and if we do not

Gualala, CA

Human impacts on the environment along our coast. The upper photo is of a select cut area recently logged for timber. Not many trees are left so weeds and brush are the new environment for the next 60 years or so. The lower photo shows the north end of Gualala where houses have been constructed on an old lumber mill site. Old mill debris has had to be removed before as well as after house construction as a couple of slides developed.

learn to control that force (e.g. ourselves), we will degrade the environment to the detriment, if not the destruction, of us all. It is difficult to deny that we humans have had a significantly negative impact in altering the natural cycle of climate change.

I hope that as you visit our area or decide to live here (especially if constructing a new home), that you will "try to live lightly on the land" and strive to preserve and restore our stunning and incredible natural environment.

End notes

My thoughts on how to best utilize the Appendix Road Log.

From Bodega Bay to Elk, public access points are identified by a mile marker and name, that is if it has an identified name. Herein I'm providing you with a brief geologic description of elements which are present at most locations. Each spot along our coast is unique and beautiful in some sense—the point is to see, enjoy, contemplate…and maybe gain a little more understanding of the forces of nature which formed these distinctive features.

I meet and also observe many people as I walk our local trails, and I often wonder why some of them are here or what they are actually experiencing while seaside. Many of them tumble along the trails in groups, talking and laughing. Many wear earphones and are immersed in their own private audio world. Often folks walk two or three abreast and are engaged in ongoing conversations. The joggers appear focused on covering as many miles as possible on their morning or evening jaunt.

The trails are becoming wider and more manicured to accommodate the influx of visitors. Some areas may eventually be paved for wheelchair access. This progress reflects our society's commitment to everyone having access to the coast.

At the same time, we want to protect this environmentally-sensitive region. Erosion threatens trails and homes located near the bluff edge. As a result, much of the California bluff has been armored with concrete and rock to reduce erosion and curtail bluff retreat. Jetties have been built to keep sand on beaches and protect marinas. The late David Brower, one of our most noted environmentalists, railed against this philosophy along with many others. Brower instead believed we should do little or nothing, and just let nature have its way. He asserted we shouldn't create wide trails, and reasoned that not everyone needs to access particularly susceptible areas. Overall, his concept was that possibly nature will be best preserved if human access is limited, or even prevented altogether.

Personally, I would suggest visitors leave the fast-paced city behind, take off your earphones, shut off your cell phone, and don't call or text anyone while strolling along this peaceful terrain. Slow down…and walk quietly in single file. This approach may lead you to enjoying a whole new level of experience as a result. We all come to the coast to get away from the bustle of our too-busy, digital lives—so why carry this intrusive technology with you? In the span of just a few feet of quiet examination and contemplation, you may find yourself feeling more invigorated than if you'd jogged three miles with your headphones drowning out the ocean's methodic lull.

This book is an exploration of the geology along the coast, but each area is also home to unique floras and faunas, so more questions are spawned as one examines further. Why do seals live on Green Cove Beach and not a quarter mile down the coast? How do these fragile wildflowers live on rocky, windswept Black Point—especially as people can hardly stand upright at times in the fierce wind there? Why are there abalone here and not someplace else? Why is this a rock beach rather than a sand beach? Why are the starfish dying? What is the name of that purple flower? What kind of hawk is that?

Take your time and explore one small area at a time. The comments I've provided will help, but you will surely have your own observations as well. Our environment is dynamic and changes, even as we watch it—and it especially changes over time. Adopt your own special spot and watch it and protect it over the long-term as you return to the coast on future visits.

Happy trails…!

Bluff Trail above Black Point Beach, The Sea Ranch

Source: Mike Lane Graphics

Geologic Road Log ~
traveling along California's scenic Highway 1

APPENDIX

Geologic Summary

Wow…we're finally here!

From our studies, we note that this small piece of land that we live on has had a rough and bumpy trip from its beginnings in southern California. This geologic journey will continue for millions of years until we plunge into the abyss at the triple junction off Cape Mendocino.

We must stretch our imaginations back to the prime event that began the forces which shaped our local geology. Let us go back to the Permian Period, 250 to possibly 300 million years ago. All the continents were clustered together in one giant continent, now known as Pangaea. The plutonic gods lay hidden in the earth's mantle, under the supercontinent. The gods became angry and their debate grew hot. Pangaea was split apart in the following Triassic Period and the continents fled from one another. North America left Old Europe and Asia, and headed northwestward. South America split to the southwest. Africa went to the south, and Antarctica fled to the far south to cool off.

The west coasts of the Americas pushed over the ocean plates and subduction became the new name of the game. Whatever old sediments existed on the west coast of California were shoved under the continent by subduction. Erosion from the continent spilled sediments into the sea, west of what are now the Sierras. The sediments were named the Franciscan Formation. A coastline was formed. The dinosaurs developed on the continent, but the Franciscan Formation area was under the ocean, so only experienced marine forms of life. The continent kept moving westward and the Franciscan Formation was crumpled, squeezed, and faulted, later thrust upward to become the Coastal Ranges. Some areas were squeezed to the point that the rocks suffered low-grade metamorphism.

The struggling plates off the west coast continued to plunge under the continent, and the Farallon Plate was consumed by the subduction. Complete melting of the rocks occurred. Subsequently, volcanoes and lava flows developed 60–100 miles inland from the coast. After the Farallon Plate was consumed, around 25–29 million years ago, either the northern drift of North America slowed, or the speed of the Pacific oceanic plate increased, resulting in the formation of the San Andreas transform Fault.

The offshore basins, (the Gualala Basin, the Bodega Basin, and others) had quietly existed offshore receiving sediments from the eroding continent. The change in direction of the Pacific Plate movement carried the Gualala Block on a northern trip to its current location. In the process, the Gualala Block was folded and faulted, both onshore and offshore. *Transform movements* (strike-slip as opposed to vertical movement) have been measured over a wide zone from Clear Lake and Healdsburg to several miles offshore in the deeper basins at the edge of the continental slope.

The San Andreas Fault has split a small narrow section off from our offshore basins and now we can examine the ocean sediments and rock formations from Bodega Bay to Alder Creek, just north of Point Arena. The following Appendix Road Log section takes us from Bodega Bay to Elk—and as we examine each site, we learn more still about our region's fascinating geology.

Points of interest from Bodega Bay to Elk, California

Can you really understand it unless you personally experience it?

The following suggested stops along your driving excursion begin at Bodega Bay and proceed north to Elk, covering 85 miles of our beautiful and dramatic coastline. Several side trips are also suggested to allow you to explore features located inland. Mile markers along Highway One are identified to help you locate features or indicate coastal access points when available.

Bodega Bay area

You'll approach Bodega Bay from the south, through a small canyon on Highway One named Cheney Gulch. If you have traveled up the coast from the Bay Area, you will have likely noticed a change in topography and a change in vegetation on this drive. Much of the region between Petaluma is primarily grassland and the way to the coast is through an area of broad valleys and small streams. Cattle and chicken ranches are the chief occupations on large farms situated along the highway.

Rainfall in this area is less than the 30+" per year required by a redwood forest. The dominate trees here are eucalyptus which were imported and planted as windbreaks to protect grazing cattle. Two Rock and many of the other protruding mounds poking out of the landscape are metamorphic *blueschist clinkers* in the Franciscan Formation. These were formed during intense folding and faulting of the coastal ranges.

⮑   **SON 9.0 & 9.16** are the turnoffs to **Doran County Park** which is located on the bay mouth bar which blocks Bodega Bay. The *San Andreas Fault* is poorly defined in this area. The movement along the 1906 trace was close to the eastern bluff edge near Highway One. The San Andreas Fault zone encompasses much of the bay area, which is nearly a mile in width. Since the bay is filled with mud and the north shore area is covered with sand dunes, movements during the 1906 Earthquake were easily erased following the event. The only documentation recorded of the 1906 event indicated the fault trace was located east of a house on the north end of the bay, no lateral offset was noted. Another trace or pressure ridge was situated 200 yards west of the house with 18" of vertical displacement noted. The soft sand in the dunes and the mud deposits in the bay may be subject to liquefaction during large earthquake shaking events.

Doran Regional Park, a Sonoma County campground, is situated on the bar. There are also several other campgrounds in the Bodega Bay area. As you first enter the approach to the bar, you'll cross the 1906 trace of the San Andreas Fault. Tidal marshes occur on both sides of the bar and have abundant wildlife, so this is an ideal spot for nature watching.

Just north along Highway One is **Bird Walk Dunes** coastal access with parking adjacent. Picnic tables and trails through the dunes make this a nice spot to stop for a picnic and exploration. Additional (limited) parking is also located next to the highway which provides more direct access to the dunes.

Proceeding north along Highway One, you will next see a short, paved turnoff next to the bay which offers access to lodging facilities and a few fast food restaurants. Here there are excellent views of the bay as well as the bay mouth bar to the south. This loop returns to Highway One just south of Lucas Wharf.

⊃ **SON 11.25** A **Visitor's Center** is located at the south edge of Bodega Bay on the east side of the highway near the gas station. Detailed maps and tourism literature covering camping, lodging, shops, and restaurants are available here.

Stopping at this spot will provide you with access to the bluff, harbor, and **Bodega Head**. East Shore Road descends to the ocean, then turns right onto Big Flat Road which crosses an extensive low-lying area of sand dunes. These dunes are partially stabilized with vegetation. Turning south onto West Side Road along the west side of the bay, you'll pass boat docks and a small restaurant or two. Proceed south to the marvelous Bodega Head.

There's a parking lot located just before the road climbs up the dunes and heads toward the ocean. This area, known as the **Hole in the Head,** was purchased by PG&E in the 1950s for a nuclear reactor power generating station. Construction began in the early 1960s but was defeated by environmentalists in 1964. Take the trail to the south and west and you can see the *"Hole in the Head"* which was excavated for the reactor site. Geologic studies presented at the time indicated the planned reactor was located 1000' west of the San Andreas Fault and accom-

panying seismic studies indicated the most fractured bedrock was 3000' east of the reactor site. Incidentally, a fault cut across the north wall of the excavated hole. Evidence (according to the report) stated the overlying and surrounding soil indicated the fault was old, but that movement had occurred there during the early Pleistocene.

Nearly everyone living in the immediate area as well as many people from other parts of Sonoma and Marin Counties suddenly became environmentalists and a huge protest defeated the construction of the Bodega Head nuclear power plant. Early environmentalists organized for this challenge went on to form the California Coastal Commission whose first act was to shut down development at The Sea Ranch, originally proposed with a 5000-home capacity. A compromise was reached determining a maximum build-out of 2329 homes. Another nuclear power station was proposed at Point Arena Cove in 1971, and was again defeated by local environmentalists. (*If you take this trip up the coast, even if you're a capitalist, a developer, or an oil & gas geologist, I almost defy you to not become an environmentalist too!*)

Before you leave the parking lot, take in the scenic view east to the entrance to Bodega Bay, the dredged ship channel, and the County campsite at the end of the bar.

After your visit to the Hole in the Head, proceed uphill to the west parking area next to the ocean bluff edge. The head consists of bedrock granite and granodiorite, which has been dated as Early Cretaceous, formed by the melting of pre-existing rocks approximately 100 million years ago. This granite matches the granite in Point Reyes located only eight miles to the south and across a small stretch of ocean. The San Andreas Fault enters the south end of Point Reyes as three faults: the Golden Gate Fault, the San Andreas Fault, and the San Gregorio Fault. Tomales Bay averages a mile and a half in width. Point Reyes was offset 20' during the 1906 Earthquake. It is probable that the Bodega Head was also offset by 20' during the 1906 event.

The Bodega Granite is similar to granites in Southern California, and is thought to be evidence of the amount of offset of by the San Andreas Fault. Remember that the San Andreas Fault is identified as the boundary between the North American and the Pacific Plate. However in actuality, the boundary and the plate movement are spread over several faults, rather than just along the San Andreas Fault. The granite on Bodega Head has been identified as part of the *Salinian Block* and is located west of the San Gregorio Fault. Granites on Bodega Head correlate with granitic rocks found 300 miles to the south in the Salinas Valley. Other documented movements on the San Andreas Fault produce 225 miles of offset of rocks at the Pinnacles in the Salinas Valley which correlate to similar rocks in southern California. Additional offset has been measured on the San Gregorio Fault which, coupled with the San Andreas, make up 300 or more miles of plate offset.

By the time you have read this information on-site and examined the granite and surrounding rocks, the wind will have caused you to move back toward the highway—the Bodega Head is one of the windiest spots on our coastline! There are some quiet days here however too. If you catch one of these or are warmly dressed, there are trails around the head with marvelous views of the coast.

Proceeding north along Highway One, you come to **Bodega Dunes Campground**. Again this is an area of sand dunes and coastal trails. Near the entrance on the west side of the road is an area of windblown hollows or blowouts in the dunes. Eddies of wind and the lack of vegetation or dried-up shallow water pools have caused these hollows or blowout areas. Vegetation is important in keeping the dunes stable, so it is desirable to keep to the trails so as to protect the fragile plants.

⊃ **SON 12.40 Salmon Creek Bridge** crosses a wide fault zone identified as the Salmon Creek Fault Zone, which is separated from the 1906 SAF (San Andreas Fault) trace, and probably marks an older faulting associated with the folding of the Coastal Ranges. The Franciscan rocks in this zone are identified as a mélange of faulted and churned sandstones.

The **Ranger Station** entrance is located just north of the bridge on the north end of **Salmon Creek Beach** which extends two and a half miles south to Bodega Head. The beach is a favorite spot for surfers, and on most days, you will see several in the water waiting to catch the next wave. There are several roadside parking areas with good access to the beach here. You may notice there are no sea stacks off this beach area, probably because the San Andreas Fault is located just offshore, and the sandstone bedrock is highly fractured and easily eroded. The bedrock is composed of sandstone belonging to the Franciscan Formation. Rocks strike NW–SE with dips of 60º to the northeast.

⊃ SON 13.21 Continuing north, one finds a long stretch of coastline with many scenic **sea stacks**. There are also many turnouts providing great views from your car and usually beach access as well. The sea stacks here are marvelous and very photogenic. The stacks consist of hard spots in the Franciscan Formation. During the folding, faulting, and uplift of the Coastal Ranges, squeezing of the rocks along fault zones produced *greenschist metamorphic clinkers*. The surrounding rocks are sedimentary sandstones, siltstones, and shale, with the hard pods of metamorphic rocks. Various mineral crystals, including garnets, epidote, mica and others are found within these clinkers.

Proceeding northward, scenic stops occur at the following: **Miwok Beach, Coleman Beach, Arched Rock Beach** and others.

⊃ SON 13.84 marks the position of the **Arched Rock** located just offshore. The best spots to view the arch are from south of the rock. **Carmet Beach** is adjacent to the settlement of Carmet. Growing along this section of the bluff are large areas of ice plant which, although not a native species, have taken over broad areas. Ice plant is a good stabilization vegetation in preventing bluff erosion. Additional parking areas are at **Schoolhouse Beach** and **Portuguese Beach**.

⊃ SON 15.22 **Gleason Beach** is at the mouth of **Scotty Creek**. A very narrow strip of land sits between Highway One and the ocean, and several houses have been built at the bluff edge. The underlying sandstone bedrock is very hard and, at first glance, is suitable for supporting house foundations. However, the problem is that the bluff edge is highly fractured and jointed here. The waves from ocean storms are concentrated between the offshore sea stacks at times, and thus pound the bluff edge head-on. Septic tanks were placed adjacent and behind the sea walls. Leachate and groundwater contributed to the amount plus speed of erosion of the bluff behind the sea walls. Very extensive armoring of the bluff face was constructed of concrete, but "Mother Nature" found a way to undercut the sea walls. Consequently, several houses have been removed from the bluff edge at great loss to their owners. Other houses are "red tagged" by Sonoma County for demolition, again at the owner's expense.

Bluff erosion and retreat is a serious problem at many locales along the California coast. We come to the coast during the summer season and the waves are gentle and the ocean serene. We look at a rock, pound it with a hammer, and assume it will be there forever. We think, "Ah, this is where I want to put my house!" But come to the coast during a heavy winter storm and note the difference. Then you will not only witness huge and intense waves, but the ocean is also armed with rolling rocks and sand. One storm can cut out a section of bluff as much as 10' in thickness. If the current period of climate change continues, spurring rising sea levels and more intense storms, this pattern of erosion will grow more severe. The California Coastal Commission does not like continued armoring of the bluff—the homeowner is last on their list of things to protect. Instead, the protection of cities, marinas, and public-serving facilities all come before homeowners in terms of proposed bluff stabilization projects.

Duncan's Bay and Duncan's Point both offer points of interest off the highway. Here, the road circles around some fossil sea stacks protruding above the coastal terrace. The local legend is that during the Pleistocene, herds of mammoth roamed the coastal area here. The undercutting of the protruding sea stack was said to be an area where the mammoths rubbed their backs against the rock and protected themselves from the wind. Maybe this is a fanciful tale, but it's possible. Native Americans, after their arrival on the coast 30,000 years or so ago, hunted all the large animals to extinction.

⮑ **SON 17.00 Wrights Beach** offers a campground located right on the beach. Near the entrance stands a greenschist boulder with a patch of green chlorite-bearing rock which is worth examination.

➲ **SON 18.22 Shell Beach** has a 1.5 mile trail along the ocean called the Kortum Tail. It's an interesting trail with a variety of metamorphic minerals in a greenschist clinker at the bluff edge.

➲ **SON 19.20 Goat Rock Beach** is located at the entrance to the mouth of the Russian River with a parking lot adjacent to the mouth bar blocking the river. Goat Rock is an offshore sea stack. As you drive down the road or look at Goat Rock from the pullouts north of Jenner, you'll see the top of the bluff and the top of Goat Rock form a perfect sloping line. Wave erosion since the last Pleistocene glaciation, which ended 10,000 years ago, has eroded the bluff separating Goat Rock from the shore. You can crawl over a line of rocks from the beach to the rock at low tide or low-wave periods.

➲ **SON 19.22 Russian River Bridge**. A restaurant is located at the south end of the bridge. It has gone through several incarnations and is currently a Russian restaurant. A single daily meal is offered by the hosts, two Russian women, one of whom is likely to join you at your table and a delightful discussion may ensue. You pay the bill by determining the meal's value and paying accordingly—worth a visit!

➲ **SON 20.20** Marks the junction inland on County Road 116 to Duncan Mills and further inland, or you may continue north on Highway One along the Russian River Estuary.

If you are traveling inland from here, it's 3.94 miles to **Duncan Mills**, which was once the end of the railroad line. It's eight miles from the coast to **Monte Rio**, and 12 miles to **Guerneville**. On the north side of Highway 116, for a distance of 3.5 miles, you will notice many small landslide scars. In fact, a few years ago, one slide blocked Highway 116 during a high rainfall event. Although the steepness of the slopes and the type of underlying soils were important factors in terms of the cause of the slides, I think that overgrazing by sheep was actually the major

factor. You will notice animal trails forming small benches covering large sections of the hillsides here. The surface waters from storms follow these trails, and at a downslope dip of the trails, causes water to concentrate and overflow the trail.

⊃ **SON 22.20** Marks the **mouth of the Russian River**. The two miles of estuary west of the bridge include the small hamlet of **Jenner,** which hosts some tourist facilities, lodging, vacation rentals, gas station, post office, deli, one or two restaurants (if open), and a public restroom facility. From Jenner, it's approximately 24 miles to the next gas station, deli, and lodging, with only a generous stretch of handsome scenery in between.

There are several terrific off-road spots to park so you can enjoy views of the river, ocean, and wildlife just north of Jenner. Often, there are hundreds of seals lying on the sandbar at the Russian River estuary mouth. Some nice sea stacks are located here including one with a cave extending through the stack. Sunsets are marvelous here! And this is a favorite spot for photographers and artists. The rocks change from black in the winter to white in the summer. Why is that? *(Hint: ask the birds.)*

⊃ **SON 22.96** Look north on the hillside above the road and you will see a triangular rock on top of a rounded knoll painted pink (!)—locals call this "**Titty Rock**." If you let your imagination go, you can see an additional breast just to the east and a rounded, lying-down figure with a pregnant belly to the west. The head and the hair (a patch of brush) come down to the highway. *Sometimes one has to look for these features from both directions, and possibly with a dose of creativity. You be the judge!*

⊃ **SON 24.42** This spot offers a good view of a couple of scenic sea stacks.

⊃ **SON 24.50 Russian Gulch** has a parking lot just across the stream on the north and west of the highway. These rocks consist of Franciscan sandstones. The path here offers a nice stroll to the ocean so this is a popular spot to take a hike and snap some photos.

⊃ **SON 25.53** Parking is available on the east side of a switchback near the top of the grade. There are terrific views down the coast toward Jenner from the west side of the road. Watch for traffic when crossing the highway here.

⊃ **SON 26.19** Turn left into the **Vista Trail Access**. There is a public restroom here and a paved trail with wheelchair access along this high bluff area. The elevation is about 600' and the trail circles to the top of the hill with expansive views south, east, and west. This is a must-see spot!

A pleasant detour is **Meyers Grade Road** which is the next turnoff on the right. It climbs from 600' elevation to the top of the ridge at over 1100'. Great views of the coast are looking back toward the south, but there is only one spot to stop for viewing traveling north, so the best views will be enjoyed from this area when travelling south. Next to the road on the east side are numerous blueschist knobs which rise up from the slope. Three miles further is another patch of blueschist knobs, and these decrease in number as one travels further north. Published geologic maps indicate these knobs end abruptly in a line straight to the east. This may indicate that the Franciscan Formation has been less squeezed and not metamorphosed further north compared to the region you've just traveled through. I suspect this is explained by a regional cross fault cutting the coastal range and marking the northern boundary of the knobs.

Meyers Grade Road runs 9.0 miles to **Timber Cove Road** which will bring you back to the coast, the entire loop is 11.0 miles. Along the way you'll pass the Fort Ross Winery, an art glass studio, a house which was once an old stagecoach stop, and the Fort Ross School. At 9.4 miles along Timber Cove Road (which is located only 1.4 miles east of Highway One), the road levels off and takes a switchback. This is the 1906 trace of the San Andreas Fault.

If you don't take the loop but instead stay on Highway One, the next few miles offer magnificent ocean views with numerous convenient parking spots.

- **SON 27.05** Marks a slight draw, which is a probable fault zone and the last blueschist knob located in proximity to the highway.

- **SON 27.74** Identifies the site of an old landslide which kept Caltrans busy for several years as, each year, more material kept sliding onto the road here. This slide area seems stabilized with vegetation currently. The rocks consist of grey-green, very hard, fine-grained sandstone shot through with quartz veins. It is almost a meta-quartzite. Quartz veins are a good indicator of faulting.

- **SON 28.63** This spot encounters some very hard crystalline quartz sandstone with very steep bedding. To the north of this spot, the slopes are mostly brown sand and soil with sandstone fragments.

Stop on the left just before the roads snakes around **Timber Gulch**. Across the gulch you can see a Caltrans concrete retaining wall sculpted to look like rock. Some of these early walls allowed hand holds for rock climbers, but now Caltrans walls are designed to prevent climbing.

The grey rock in the gulch is Franciscan Formation and fault gouge. The rocks on the left (west) are Tertiary sandstones belonging to the Galloway Formation which underlie coastal Terrace I. The Galloway Formation is approximately 120 million years younger than the Franciscan Formation. The contact between the two formations is the San Andreas Fault.

⏵ SON 29.90 Marks the **Timber Gulch Stream** Trace.

⏵ SON 30.20 Caltrans faux rock wall. Landsliding and slumping of highly fractured bedrock and breccia have plagued this area along Highway One for many years. The wall was constructed to protect this section of the highway. The concrete wall is an attempt by Caltrans to make the wall appear natural—does it look natural to you? *(Not to a geologist! Maybe it's better than just raw concrete though.)*

⏵ SON 30.30 This is where the San Andreas Fault comes back onshore. It's difficult to see more than one trace at this location, but I suspect at least one additional trace at the notch in the Caltrans wall. Less than a mile north, we identify four traces from the 1906 Earthquake event.

⏵ SON 30.75 **Mill Creek**. USGS studies have examined the San Andreas Fault offset movements at Mill Creek by trenching across the fault and determining ages of the movements. You can notice the offset of the stream from the highway. If you look to the east and north of the creek, you will see a large landslide which blocked the creek and caused Highway One to wash out in the 1970s. A one-lane metal Bailey bridge was trucked in to span the void which which we used for more than a year until the road could be rebuilt.

⏵ SON 31.00 The white Pedotti Ranch Buildings are grouped on the west side of the road. A few years ago, a tornado touched down next to the road and completely destroyed the barn closest to the highway. I found it hard to believe we had tornadoes around here—I thought I had left them behind in Oklahoma! In examining the debris remains of the barn, the broken boards were scattered on all sides of the demolished structure on both sides of the highway, some landing up to a couple hundred feet away. Some of these boards appeared to have fallen in a rough, circular pattern.

⏵ SON 31.37 Entrance to **Reef Campground** and trail to the ocean begins here. This ravine appears to follow a crack of the San Andreas Fault. Caltrans recently rebuilt the road and bank at the north end of the ravine, just north of the switchbacks.

⊃ **SON 30.75 to SON 31.76** Look at the hillside to the east. It is a massive series of landslide and slump materials. At one point, near the campground entrance, the slide is raising the road as it slowly creeps downslope. Slow down or you will feel it!

⊃ **SON 32.19** **Russian cemetery** for Fort Ross. A few years ago, the cemetery was scraped clean of vegetation, which then revealed the outlines of the graves. Several graves were excavated and documented for remains and artifacts. Wood markers rise above some of these gravesites. Details of the excavation and other cultural information are available at the Fort Ross bookstore.

⊃ **SON 32.45** **Fort Ross Creek** is offset a half mile to the east along the San Andreas Fault. An old map indicates the Russians mined coal for fuel along this creek. No one has found any traces of coal in the streambed recently. It's possible the movement of the fault may have squeezed some wood into a low-grade coal deposit. If it was mined, it probably was of inferior quality and was mined out.

⊃ **SON 32.93** Entrance to historic **Fort Ross** (opened in 1812, originally known as Fortress Ross). Visit the Fort, the surrounding McCall Farm buildings, and explore marvelous coastal features hereabout. Next to the parking lot is a paleo-sea stack protruding above Terrace I. An easy climb to the top gives one a chance to spot migrating whales and other wildlife. Fort Ross is a favorite tourist stop and revisited often by local inhabitants as well, and there are many interesting events scheduled there throughout the year. The local Kashyia Indians, visiting Russians, and local historians often have public gatherings at the fort as well.

Across Highway One, turn east on **Fort Ross Road** and proceed uphill for a half mile to where the San Andreas Fault crosses the road with four traces. To the south is an orchard with several picnic tables. The depression to the east is a sag pond. There's limited parking across the road at the entrance to **Stanley S. Spyra Memorial Grove.** Park here and take the trail to the north for excellent views of sag ponds, pressure ridges, trees with broken tops, and other fault-related features. Slow creep of the faults has caused the tall redwood trees to twist and slant in various odd directions here. Adjacent to the road is a huge Bay Tree which splits into many trunks and is over 40' in circumference. Further along the trail is a double redwood ring of 18 trees, several with tops snapped by earthquake shaking. *This hike is a must!*

⊃ **SON 34.20** There is a small rough trail to the base of **Kolmer Gulch** with space for one or two cars on the west side of the road. Parking should face south.

⊃   **SON 34.36 Kolmer Gulch.** Massive sandstone beds outcrop at the ocean. The San Andreas Fault is three-fourths of a mile inland. Fault line tributaries extend north for a third of a mile and south for more than a mile along the 1906 fault trace.

⊃   **SON 34.60 Windermere Point, Fort Ross State Historic Park.** Not marked but a good exit is located at the north end of the road curve out of the gulch. There's a nice parking area near the ocean just after the curve, just south of the Fort Ross Lodge. **The Fort Ross Store** is on the right side of the Highway and has a deli and gasoline. The next gas is approximately 15 miles north at **Stewarts Point Store**. From the point stop, look north across **Timber Cove** to the **Timber Cove Inn**. These rocks are sandstones of the German Rancho Formation and dip 45° to the northwest. The view to the south is of a small cove with a sandy beach. Faulting in the cove has created a jumble of large sandstone blocks sliding into the cove.

⊃   **SON 35.44 Timber Cove Boat Landing.** There are invariably several trailers or campers situated here, along with a ramp to the ocean.

⊃   **SON 36.60** Brings you to the **Timber Cove Lodge** which is a marvelous building with a massive stone fireplace—a good spot to get warm after poking along the coast, especially if you're traveling here in winter. Lunch with a drink from the bar also helps. The massive totem pole sits on a paleo-sea stack and is accessible from inside the lodge. The view of the coast, the totem pole, and the surrounding shaped sandstone boulders are definitely worth a visit.

⊃   **SON 36.00 to 37.00** is the **Timber Cove private residential community.**

⊃   **SON 37.13 Stillwater Cove Regional Park.** There's a campground on the east side of the highway with a trail to the ocean and a public restroom. The rocks are massive sandstones falling into the ocean. Parking is limited along the highway here. On the north side of the gulch is another trail to the beach at SON 37.52.

⊃   **SON 38.00 Ocean Cove Store and Campground.** This is a popular private campground along a scenic stretch of coastline.

⊃   **SON 38.24 Ocean Cove Lodge and Restaurant** located on the east side of the highway. A great place to eat and enjoy the sunset.

⊃   **SON 38.37** is the south entrance to **Salt Point State Park.** This park stretches along the coast for six miles, covering 6,000 acres with several good coastal access points. It's a favorite place for gathering wild mushrooms and for abalone picking. There are two campgrounds here. On the east side is **Woodside Campground.**

⊃   **SON 39.89 Gerstle Cove Campground** is located west of the highway. Along the bluff front, you'll notice an interesting erosion pattern known as **tafoni** as the sandstone surface is weathered into an intricate pattern of circles and holes. Sea Urchins can also bore into rocks leaving round, regular holes. These holes are different in that the pattern is more irregular and the holes develop on the bluff face, above the normal tide levels. It is thought the saltwater which penetrates the sandstone rock leaches out the rock cement, and wind and waves then do the rest. You may note a dark or orange-colored stain around the holes produced by the concentration of iron minerals here. These tafoni are also prevalent at Fisk Mill Cove and many other spots along the coast.

⊃   **SON 40.80** is a large parking area on the east side of the highway. There are many paths across the wide meadow to the ocean here. At the north end is a spot where storm waves crash dramatically, providing a spectacular photo opportunity.

⊃   **SON 41.00** Marks the south stream of the two streams which flow into **Stump Gulch.** The highway makes a sharp bend across the stream here. The vegetation east of the highway has two tiers sloping toward the ocean and is a favorite view spot.

⊃   **SON 41.22 Stump Beach** is a great stop for a break as there's good parking, five picnic tables, and a public restroom. The graveled trail is worth the short trip to a small cove. The bluff is nearly vertical consisting of thin-bedded sandstones and siltstones, and with a low dip into the bluff. Sea caves are accessible from the tan-colored sandy beach. Two small streams join together at the ocean edge. The rocks to the south are massive sandstones with steep bedding dip. The formation of the cove appears to be fault controlled, with a bounding fault on each side of the cove.

⊃   **SON 41.35** The highway crosses another small stream here at **Chinese Gulch.**

⊃   **SON 42.63 Fish Mill Cove** hosts another paved coastal access path on the north end of Salt Point State Park. There are two parking areas north and south of the access.

⊃   **SON 42.70 Kruse Rhodendron Park** is located to the east on **Kruse Ranch Road.** This road is mostly gravel and steeply climbs the coastal ridge to Tin Barn Road which joins Seaview road, travelling south. The park is comprised of 317 acres and has three miles of hiking trails. When in season, the flowers cover a large area of the forest here.

Proceeding north along Highway One for nearly a half mile is a long stretch of eucalyptus trees.

⊃   **SON 43.06** A good spot for parking west of the highway just before you reach the remains of the old Kruse Ranch buildings. This is the south end of the **Stewarts Point Marine Reserve.** There is a trail to the ocean which leads to **Sentinel Rock.**

⊃   **PG&E Marker 45.35 Salt Point State Park.** Long trails circle the meadows around paleo-sea stacks on the coastal terraces and descend to the bluff edge. This is a very quiet area blocked from road noise which allows one to contemplate nature and enjoy observing birds, flowers, and other wildlife.

⊃   **SON 44.63** Rough trail (not recommended) to the coast at the north end of Salt Point State Park.

⊃   **SON 45.1** Call Box parking area. Nice view of a cove to the south. Terrace I and Terrace II can be easily identified here sloping toward the ocean.

⊃   **SON 47.88** Remains of the **Stewarts Point School**. Located on the south side of **Fisherman's Bay.**

⊃   **SON 48.01 Stewarts Point**. Many of the buildings are historic structures dating back to the 1800s when Stewarts Point was an active dog-hole schooner seaport (this was a common name given to small, rural ports on the Pacific Coast between Central California and Southern Oregon which operated between the mid-1800s until the 1930s). There is a store currently operating, deli & restaurant/bar, gas pumps, and a post office. Adjacent buildings consist of numerous barns and cottages as well as a residence which was once a hotel. Historically, the post office moved back and forth from Stewarts Point to Black Point a couple miles to the north. The **Stewarts Point Post Office** is a USPS branch serving the region, including the **Kashia Band of Pomo Indians** reservation and surrounding area. (The Sea Ranch Post Office is a private contract post office at the Sea Ranch Lodge.)

The coastline here is not currently open to the public; although a conservation easement covers some of the area, and it will be accessible at some future time.

Skaggs Spring Road is across the road, east of the store, and heads inland to Lake Healdsburg and Sonoma. The road turns abruptly south 8/10 of a mile east from Highway One. This marks the trace of the San Andreas Fault Zone. Approximately 6/10 of a mile further, the road takes an abrupt turn east across another trace of the fault.

➲ SON 50.59 **The Sea Ranch Lodge** offers lodging, a restaurant with full bar, a gift shop, and a post office. It is the gathering place for locals as well as tourists. Whale watching can be enjoyed from the comfort of the restaurant or solarium. The Lodge grounds have become a popular place for weddings and reunions. There is a marvelous historic barn located at bluff edge. This was once an extensive commercial area with a dog-hole port and 13 adjacent buildings. The barn at bluff edge and a small sheepherders shack are all that are now left of the original buildings.

There is a **Tesla charging station** and an additional charging station for other types of electric vehicles here.

➲ SON 49.70 to 58.48 **The Sea Ranch.**

The developer of The Sea Ranch, Oceanic Properties, Inc., built the lodge, post office, store, restaurant, and bar between 1965–68. The Sea Ranch was previously a 5,000 acre sheep ranch, and is now a 3300-acre planned community. The remaining lands were sold for timber harvest. There are currently 1800 homes completed with a maximum potential of approximately 2200. Recreational facilities include tennis courts, three swimming pools, meeting rooms in two of the original (remodeled) houses, horse stables, an 18-hole public golf course, and the Knipp Stengel barn (built in 1880) which is home to the Sea Ranch Thespians and their theatrical productions. Also housed here is The Sea Ranch offices, maintenance barns, water company office, two fire stations, a commercial area of small offices and a deli (Two Fish), storage units, and a building supply company.

➲ SON 50.85 **Black Point Beach Public Access** and parking lot are located just north of The Sea Ranch Lodge. Black Point and Bihlers Point are among the best places along the coast to view wildflowers and a flower trail map is available at The Sea Ranch Lodge. It always amazes me that seemingly delicate wildflowers can grow in such abundance on this windy landscape. You'll see the best variety in the spring, but something is always blooming thereabout.

Follow the trail to **Black Point Beach**. The beach is shielded from the wind more here than at the top of the bluff. Black Point beach is the longest beach on The Sea Ranch. At high tide, especially during winter months, the waves may break into the bluff, so it's advisable to keep an eye on the ocean. There is only one stairway access to the beach, located near the south end. The bluff is nearly vertical due to erosion of the hard resistant meta-basalt known as the **Black Point Spilite**. (See page 22 for more detailed information.)

On the south end of the beach is the fault contact of the Black Point Spilite with the Stewarts Point Member of the Gualala Formation. The fault zone is 12' or more in thickness, and consists of black shale and broken rocks (a breccia). The Stewarts Point Member is a massive conglomerate with large cobbles of quartzite, granite, and other rocks. Much of it is stained very dark with iron cement. During the winter, the beach is mostly black and green in color as a byproduct from the eroding spilite. Sometimes you can see thin, elongated patterns of black sand left on the beach by the retreating waves. My handy magnet has detected that a substantial percentage of the black sand consists of magnetite crystals. Five waterfalls spill over the bluff edge during the rainy season and add to the dramatic beauty of this beach.

⮕ **SON 51.64 to 51.75** Highway pullout is located at a sharp bend in the road. Black Point Spilite forms the hill on the east side of the highway here. The formation extends under the coastal terrace deposits of sand and gravel to the bluff edge. The rock is less green than on the beach and takes on a reddish color from iron oxidation.

⮕ **SON 51.94** **Annapolis Road.** This county road also connects inland to Lake Sonoma and Healdsburg. The San Andreas Fault is located four to five-tenths of a mile to the east. The road takes two dips in fault traces as it descends downslope to the South Fork of the Gualala River. The dips are sag ponds with adjacent pressure ridges formed during the 1906 Earthquake event. The river marks the traces of older events. Two bridges cross the South Fork and Wheatfield Fork of the Gualala River at this area known as **Valley Crossing**.

⮕ SON 52.32 **Pebble Beach Public Access** is the second public access point on

The Sea Ranch. Five public accesses were established along the 10 miles of The Sea Ranch in 1981 with the passage of the Bane Bill following a legal battle and five-year moratorium on the building of new houses. The Sea Ranch was limited in size to 2329 lots from the original 5000 lots proposed by the community's developer. (Although some of us were unhappy about not being able to build our dream home during the moratorium, the decrease in size has ultimately had a positive impact overall.) Now, there are over 1800 homes constructed, of which over 500 are available via vacation rental programs. The vast unwashed "public" who were so feared can now rent a vacation home here, complete with access to over 50 miles of trails as well as The Sea Ranch's recreation centers. Californians and visitors from around the world seem to have become more respectful of our natural environment in the last couple decades, so rarely does anyone leave trash on the beaches.

This access connects to **Pebble Beach**, which, historically, was one of the best places for rock-picking of abalone. Over-picking has greatly diminished their numbers locally, but other areas are still good for diving. The rocks in the bluff consist of a series of squeezed shale, with four or more faults detectable. The Black Point Spilite extends to the south and is marked by a thrust fault dipping 45° to the southwest. The rocks on the north are also part of the Stewarts Point Member and are located on the north side of the Black Point Anticline. (Remember that the rocks on the flanks of an anticline dip in opposite directions off a core rock.)

⊃ SON 53.96 **Stengel Beach Public Access**. The parking area for this access exists just north of the Knipp Stengel Barn. This barn dates from the late 1800s, was bounced off its foundation in the 1906 Earthquake, and thereafter jacked-up and restored. When Oceanic Properties acquired the property in 1964, it was used as a horse barn for Sea Ranch members and guests. Hay bale concerts were popular at the time. Following the settlement of the Bane Bill in 1981, the developer wanted to tear down the barn and develop houses in the pasture west of the barn and extending to the bluff edge. Three Sea Ranch horse owners were incensed and filed a lawsuit against the developer. The result was that new stables were constructed just east of the highway.

The Sea Ranch Design Committee took an active role in the planning of the developed meadow west of the barn. Susan Clark was successful in placing the Knipp Stengel Barn on the National Register of Historic Places and the old barn was saved. The Friends of the Barn, a group of volunteers, contractors, and contributors banded together to restore and stabilize it structurally: a concrete foundation was constructed to replace the redwood piers which had simply sat directly upon the dirt. Restoration included a new roof, electrical panel, sprinkler system, repairing/replacing siding, and restoring windows at both ends—all of which was accomplished primarily by the volunteers working on Saturdays. These weekly work parties became a terrific social event, but they also got the job done. The barn has become one of the most used plus beloved structures on The Sea Ranch. The Friends of the Barn still meet there on Saturday mornings for coffee and they also design and construct sets for plays put on therein several times a year.

Stengel Beach marks the geological contact between the Stewarts Point Member and the Anchor Bay Member of the Gualala Formation. On both the north and south sides of the beach is a conglomerate and, at first glance, these appear quite similar to each other. However the Anchor Bay Member conglomerate is darker in color with more mafic cobbles and less quartzite. The sandstones contain more plagioclase feldspar. Check it out (in spite of the fact you may need to be a geologist to really see the difference!). However it's a lovely beach and you can also enjoy viewing the seals who like to lounge in the sun and snooze on rocks at the south end of the beach here.

⊃ **SON 55.26 Shell Beach Public Access.** This beach is one of the few on The Sea Ranch where you can find seashells. In most areas, the strength of the waves tends to breakup the shells in the surf before they reach the beach. Shell Beach is a very rocky area, so you might not expect to find shells preserved here. However, the offshore line of rocks offers protection for the shells once the waves carry them ashore.

This is a pupping area for seals in spring, and you can see numerous mother seals with their pups sunning themselves on the rocks then. During the pupping season in March and April, you'll need to stay away from close contact with the seals—especially as the pups are cute and young children want to run up close to them. The Sea Ranch has a volunteer docent group who monitor the area and politely educate visitors. Some of these docents can tell you grand stories about the seals. One of my personal experiences was as follows. Maybe 10 mothers with their pups formed an elongated circle in a shallow neck of water on the north side of the beach. After taking up to 20 minutes of barking to get their small herd in position, they all then dove in at the same moment on their quest for fish. Then it started all over again, followed by another cycle. What took them so long to get into position? Did they actually catch any fish? Only the mother seals know!

⊃ **SON 56.93 Walk-On Beach Public Access**. This trail provides access to Walk-On Beach and extends to the north along the bluff for 3.5 miles to the **Gualala Point Regional Park**. In five locations north of Sea Pine Reach the surveyed and recorded trail has eroded, which technically prevents the public from using the trail. However, The Sea Ranch Association has granted a license across Sea Ranch Commons to allow for continued public access. Erosion of the bluff is an ongoing problem both for the trails as well as for homes located close to bluff edge. Some areas are more susceptible to erosion than others. Bluff retreat ranges from zero to as much as a foot in a year. In wet years, high surf can cause weak areas to slide or slump into the ocean. A rising sea level may exacerbate this problem.

Walk-On Beach is the second longest beach here and its bluff is much lower than the rest of The Sea Ranch. Shale and siltstone rocks make up the bluff edge and belong to the German Rancho Formation. Walking south on the beach, you will encounter the rocks folded into a small anticline. Further along the beach is an area of large logs which washed up on the beach and were thrown up onto the bluff. Most of these logs are weathered and partially rotted; attesting to how many years it's been since whatever event brought them here. At the south end of the beach is an area of stabilized sand dunes along the bluff edge. Sand dunes are of course windblown deposits of sand. The grains are small, rounded, and nearly identical in size. If you look at these grains under a microscope or with a handheld lens, you will note they are frosted. An easy way to tell if the sand is beach sand or dune sand is simply walking on it. Dune sand is very soft and difficult to walk on.

The Public Trail extends north along the bluff for 3.5 miles past the end of The Sea Ranch and onto the **County Regional Park**. Most of the bluff edge has very limited beach access as these beaches are small and very rocky. Several areas have sheep fences for protection. Remember, several of these areas are rapidly eroding. The coves here are small and usually bounded by one or more faults. We think of coves as protected places from the wind, and of course they are. But the reason a cove is present means the rocks were softer, more faulted, and weaker than the surrounding areas, and thus more susceptible to erosion.

◌ SON 58.03 **Salal Creek Trail** follows along the creek to the ocean and connects north to the Gualala Point Regional Park and south as a public trail to Walk-On Beach. The trail head begins at the park's entrance and travels south next to the highway to marker SON 58.03. The walk along Salal Creek is covered with vegetation—watch out for poison oak here! This is one of the few streams in the area which cuts through the coastal terrace deposits to the underlying bedrock and this stream flows year-round. Massive blocks of sandstone rest at the stream mouth. Note how faulting has them dipping in different directions. The sandstones are massive and, interestingly, have been eroded by the wind into many unique shapes. Some look like sheep, one like a camel, many look like gargoyles, seals, birds, or whatever your imagination might suggest.

Sea Ranch Golf Links. Entrance is just south of the County Park. The course is an 18-hole public course with several holes enjoying lovely ocean views. Wildlife including birds are plentiful on this Scottish-style course. Watch out for the 13 deer who always seem to be present on the fairways!

⮕ SON 58.48 North end of Sonoma County. Entrance to **Gualala Point Regional Park**. Camping and river access are on the east side of the highway. The road to the campground was the road to the Gualala Bridge historically as the old bridge site crossed the Gualala River at a spot near the Gualala Arts building. On the west side of the highway is where the park entrance is located. Here there's ample parking and a small visitors center with rest rooms. Hiking trails to the estuary and river mouth proceed from the parking lot to the ocean. A large bay-mouth bar blocks the river for most months of the year. Storms seem to pile the sand higher each year here, but seawater also washes over the bar bringing saltwater into the lagoon. Birds abound in the estuary and on the bar. If you're lucky, you'll see a pair of river otters looking for fish and trying to sneak up on a seagull or a pelican in this spot. A good place to observe the lagoon is from the **Breakers Inn** located across the river in Gualala.

Washover fan into Gualala River Estuary

The Sonoma/ County line

is located in the center of the Gualala River Bridge

⮕ SON 58.50 & Men 0.0 **Gualala River Bridge.** Boating access to the river is by permit through Gualala Redwoods at the north edge of the bridge. A kayak rental business (AdventureRents.com) actively uses this access and they may be contacted for river access.

Old Stage Road, County Rd. 501 turns eastward just south of Gualala. The road proceeds approximately 15 miles north along the first coastal ridge, joining Iversen Road, and ending at Point Arena. The road follows the Gualala Ridge Fault which is an older trace of the San Andreas Fault. Sag ponds occur just west of the road and are viewable at Bower Park.

Just after the turnoff on Old Stage Road is a turnoff to the right, the Gualala Road.

This will take you to the Gualala Arts Building and on to **Gualala River Redwood Park Campground**. Proceed to the end of the paved road to the Green Bridge. This is the junction of the North Fork and South Fork of the Gualala River, and the trace of the San Andreas Fault. The private graveled roads extend into timber logging lands and a couple of residences.

Gualala. Gas stations, grocery stores, restaurants, lodging, shops, art galleries, post office, service businesses, hardware stores, and churches make Gualala the center of commercial and social life for the area. This small town's situated on the lower coastal terrace, overlooking the Gualala River estuary and the Pacific Ocean.

↪ MEN 0.0 The **Gualala Bluff Trail** extends along the bluff edge from the Seacliff on the Bluff, south near the Surf Motel, behind the Breakers Inn (the trail is owned and managed by the Redwood Coast Land Conservancy). Nice coastal views of the Gualala River Estuary. This trail is a great spot to view washover fans during high tides. Many of us like to check the river daily to see when the bar breeches. The lagoon may drain in a very short time during one of these breeches and the steelhead younglings then head out to sea. The returning mother fish swim upstream to spawn. The birds collect for a meal of whatever is passing by. In the old days, the fishermen collected at the mouth to catch the steelhead. The Gualala River was Jack London's favorite steelhead stream reportedly. The Gualala

Hotel used to have lots of photos of fishermen sporting strings of big fish. Now if you can catch one, you have to abide by the catch-and-release policy. In spite of restoration efforts on the Gualala River, the fishery continues to decline here, probably as a result of logging in years past and climate change now as the streams are not as cold as they once were.

↪ MEN 1.74 Steep ravine with a trail to the water. There is no parking at this site. Nice sea stack, locally referred to as **Castle Rock** which is composed of horizontal thin-bedded sandstones and shale. The Old Milano Hotel, which, sadly, had a fire a few years ago, overlooked this section of the coast. Old Coast Road #301 accesses several homes at bluff edge with excellent views to the south.

◌ **MEN 2.43** A small parking area north of town opens to an ocean view. Highway One has been moved into a small jag to the east, near a washout along the coastal bluff. Through the years, there have been many spots where erosion has caused the highway to be shifted.

◌ **MEN 3.14 to 2.76 Cooks Beach.** CR 526 is an old coastal road. There is only parking for one or two cars at Bonham Trail to Cooks Beach (the trail is owned and managed by the Redwood Coast Land Conservancy). This cove is a long, scenic, sandy beach and worthy of a stop, provided you can find a parking spot.

Pirates Cove Drive is located across from St. Orres restaurant and lodging. This is a private development with no access to Cooks Beach.

◌ **MEN 3.93** Rough trail access to the ocean.

◌ **MEN 4.35 Anchor Bay** is a small residential and commercial area. In town, are small businesses, two restaurants and a coffee shop, a real estate office, a grocery store, a laundromat, and a medical marijuana shop—and both sides of the main street are owned by one family. Although the commercial buildings are located near the ocean, there is no view. It has been suggested the town is turned backwards!

◌ **MEN 4.35** at the north end of Anchor Bay is a trail to **Anchor Bay Beach**. There is a $3 fee for visitors payable at the Anchor Bay Campground, but it's worth the cost.

◌ **MEN 4.64** The **Anchor Bay Campground** is a private campground, but allows beach access for campers and other visitors. The beach is marvelous here and the kids can build sandcastles and play in the surf (despite the chilly ocean temperature). The rocks are massive sandstones to thin-bedded siltstones, sandstones, and shale. The rocks at the south end of the beach dip 46º to the north. These are jointed and fractured, and shot with quartz veins. In some areas, the shale is squeezed and, again, shot with quartz. There are some nice turbidite flow beds. The over-steepened bluff is experiencing small slides, and debris cones line the bluff which will be removed by high tide during winter storms.

◌ **MEN 5.50 & 6.0 Fish Rock** lies to the west offshore. If you pull into the parking area and turn off your engine, you may hear a colony of sea lions barking and arguing over their spot on the rock. There is no access to the coast here. I guess some other mammals feud just like us humans!

Havens Neck is just north of Fish Rock, but has no public access and is difficult to see from Highway One. It's a prominent sandstone spire, marked with a fault that runs parallel to the ocean.

⮕ MEN 8.21 **Sail Rock** lies just offshore as a prominent spire. It's very black except for a capping of white Cormorant excrement. At the shore is a nearly matching spire of yellow-orange sandstone belonging to the Miocene Schooner Gulch section of the Galloway Formation. Sail Rock is the most southerly outcrop of the Iversen Basalt. No beach access.

⮕ MEN 9.76 **Iversen Point Road** leads to a private development of six homes with terrific south-facing views of sea stacks, and the contact of the Iversen Basalt with the overlying Galloway Formation. The rocks strike to the northwest and dip 60° to the southwest. One of these houses is available as a vacation rental. The Iversen development has private access to the beach.

Stop at **Call Box MC 99** for a good view to the south of the two formations contact. Looking to the north is a sea cave developing along a fault. There is an unofficial steep trail to the south to the ocean—I do not recommend you try descending it without a rope.

⮕ MEN 10.0 There is parking along the west side of the highway for entrance to **Hearn Gulch**. Sea caves, faults, and the contact of the Iversen Basalt with the Galloway Formation occur in a nicely protected cove ideal for a secluded picnic.

⮕ MEN 10.08 This is also part of the **Hearn Gulch** stop which was established by and is managed by the Redwood Coast Land Conservancy. Trails circle the bluff edge for views in all directions with an excellent trail to the beach.

⮕ MEN 10.32 There's a parking spot on the west side of the highway. Look past the highway marker to the south toward Iversen Point—on your left (east) is yellow-brown sandstone, and to the right (west) is the black Iversen Basalt. Walk to the south for good views of two sea caves which cut the Iversen Basalt in the small cove and extend through the bluff to the ocean.

⮕ MEN 10.50 Here is a larger off-road turnout with lots of parking. Notice the nice views to the north of the bluff edge—Monterey Formation dips steeply westward on the near bluff. The farthest view makes the Monterey Formation beds appear nearly horizontal. At the south end of the turnout road, a cove is visible to the south. The black rocks in the cove and bluff face are igneous Iversen Basalt.

There is a small pullout just south of the Schooner Gulch Bridge on the west side of the highway. This has a good view of the bluff edge to the north. At very low tides you can see the patch of bowling balls in the far distance near the ocean's edge.

⊃ **MEN 11.35 Schooner Gulch curved bridge**. A large fault down to the north, cuts off the Iversen Basalt which outcrops on the east and south side of this bridge. This formation dips steeply to the west in a narrow band approximately 1000' wide and is traceable onshore for 1.8 miles to the south.

⊃ **MEN 11.35 to 11.41 Schooner Gulch State Beach** has a parking area on the west side of the highway and nearby trails lead to the mouth of Schooner Gulch, as well as a northbound trail to **Bowling Ball Beach**. The bowling balls are only visible at low tide, however many other interesting features can be accessed at most times along the beach. The Schooner Gulch stream mouth is accessible on the south trail through a lovely grove of redwoods, and wildflowers bloom along the stream here. Published maps indicate the Iversen Basalt crosses the gulch and ends at the north tributary of Schooner Gulch. I believe the fault trace follows the south stream trace. The stream bed carries only a few cobbles and pebbles of basalt. An outcrop of basalt nearly 1000' thick should produce a large stream load of basalt, which it does not.

South of the gulch, the light-colored sandstone and shale rocks dip steeply to the west. Several large sea caves are forming along fault zones. Quartz veins mark three or more fault zones. The major faults trend is almost directly east-west. On the north side of the stream, the major fault is marked by a grey-colored breccia. Quartz and sand-filled veins mark the side of the bluff in a pattern that looks like a jagged lightning streak. At low tide, you can get around the point to the bowling balls; the other approach is from the parking area, on a trail to the northwest across the bluff terrace.

⊃ **MEN 11.41 Bowling Ball Beach** is a favorite with many locals and tourists and is accessible from the northern trail. The descent from the bluff to the ocean is difficult at the base including a section of rope ladder, but seems to not be a problem for most people. The Bowling Balls are only visible at low tide and completely out of water only during very low tides. The beach is still present at high tides except during strong storms. (Read the section of the main text for a larger discussion of these unique ball formations, see page 71.)

⮡ MEN 12.40 Offers another alternative access to Bowling Ball Beach along a good trail, although the entrance is barely visible unless you stop at this marker. When you arrive at the beach, turn south and walk for approximate a half mile to the Bowling Balls. The bluff rock face is nearly vertical here and consists of soft siltstones, mudstones, and shale. On windy days, bits of shale and mudstone rain down on the beach, attesting to its rapid erosion. You may notice a couple of larger slides which have slid off the bluff, but wave action has quickly carried away most of the landslide debris thereafter. Compared to many areas along the bluff, this area exhibits a decided lack of faulting. In fact, there are only three faults along this entire section of bluff.

Before you reach the bowling balls, you will encounter many round boulders up to 12"–18" in diameter. These appear to consist of siltstone and mudstone, quite similar to the rest of the bedrock. On closer examination, dark spots and branching-like structures make up a considerable part of each boulder. These are worm burrows. I wonder why they occur in these cobbles—possibly they were nests where the worms collected. Worm burrows are scattered elsewhere in certain layers of rocks, but not as concentrated as they are in these cobbles.

⮡ MEN 12.89 **Moat Creek public trail** wanders along the bluff edge and the beach here is spectacular (the trail is managed by the Moat Creek Managing Agency). The rock formations in the bluff are twisted and faulted in every direction. You can see thrust faults as one layer has been pushed over another. There are anticlines and synclines, recumbent folds (which is a fold lying on its side), normal faults, reverse faults, and more. Some of the shale has been squeezed like toothpaste in certain areas. Pressure flowlines caused by the folding plate cross bed boundary lines from one bed to another. So much has happened to this small area that we poor geologists must sit pondering for hours to try and unravel its mysteries. The simple explanation for this twisted mess is that the rocks failed and were squeezed onto

the Pacific Plate by San Andreas Fault movements. *The moving rug folded and failed, and the Moat Creek bluff is the result!*

⊃   MEN 14.85   On the south edge of Point Arena, turn west on Port Road to **Arena Cove**. This is the only active port located between Bodega Bay and Fort Bragg. There are two restaurants, a lodging facility, and a couple of shops at the pier area. Several fishing boats operate out of this small inlet.

As you drive to the west on Port Road, take a look at the rock strata. Along part of the road here the rock strata is nearly vertical as faults cut the bedrock into different blocks. A large anticline is located on the south side of the cove. At low tide you can work your way along the beach to the south.

⊃   MEN 14.80 - MEN 16.00   **Point Arena** is situated on one of the westernmost tips of the country's 48 states and is an incorporated city with its own government. Shops, restaurants, delis, churches, a gas station, art galleries, and bars line the small collection of streets. Lodging is limited but the old motel is currently being renovated. Point Arena has the only high school along this long stretch of coast. Neighboring Manchester, Annapolis, and Fort Ross have elementary schools. Many parents here homeschool their children. Point Arena has a community-owned library. There's also a local pharmacy which will deliver prescriptions to your home. The Arena Theater is a community-owned movie theater. It's a host of National Theater performance broadcasts and provides live entertainment regularly. Local opera lovers (100 to 150 of us) arrive at 9:00 a.m. on Saturdays mornings to enjoy opera performances broadcast live from the Met in New York 10 times a year. The Garcia River Casino is located just outside of town which also offers live entertainment. So much for culture! What to do?

A side road to explore here is Riverside Drive in Point Arena. Turn east and then left on Hathaway Creek Road to the Garcia Indian Casino. The road follows the Hathaway Creek Fault in a small valley that runs parallel to the San Andreas Fault located four miles to the east.

⊃   MEN 15.94   Park at Point Arena City Hall on the west side of the highway. **Point Arena-Stornetta California National Monument** was designated in 2014 by President Obama to protect 1665 acres of beautiful coastline stretching north of Point Arena to Manchester Beach. Take the trail on the south side of the city hall out to the coast. The first view you will encounter is south into Arena Cove. The trail north is approximately two miles to the old Coast Guard station along spectacular scenery (and geology), and then nearly a mile further to Lighthouse Road.

(You may want to visit the northern section when coming south from Lighthouse Road.)

⟳ **MEN 17.0** Turnoff of Highway One on **Lighthouse Road** to the **Point Arena Lighthouse** located two miles north of the City of Point Arena. To see the sinkholes and sea caves (described in the first part of the book, see page 74), stop and park at the coast one and a half miles from the highway turnoff. You can walk south along the coast up to a mile leading to the old Coast Guard Station. The sinkholes start about a half a mile from the parking area. The sea caves along the path are incredible. Offshore is a small island, **Sea Lion Rocks**, with a couple of good sea caves. The rock strata on the offshore island is nearly horizontal in contrast to the steeply dipping rocks onshore.

Proceed north to the **Point Arena Lighthouse**, reconstructed after being damaged in the 1906 Earthquake. The Lighthouse is owned by a non-profit group who have restored it. There are four cottages used as vacation rentals here and the museum building offers interesting exhibits including the Fresnel lens previously used for the lighthouse's beacon. At the top of the lighthouse, docents explain the facility's prior operation and interesting history. The view is marvelous and worth the climb.

Returning to Highway One from here, we turn north at the **Lighthouse Pointe Resort** which has cabins as well as the **Rollerville Cafe**. The topographic maps indicate this was Flumeville—an ambitious project in the late 1800s which captured water a mile and a half upstream on the Garcia River. The flume powered a large, overshot waterwheel which drove a cog-like device and the hoisting works, lifting lumber 300' from the valley floor where the mill was located. It milled the lumber from logs originally floated down the river. A railroad situated at the top of the hill transported the lumber about a mile to the port where doghole schooners were then loaded from a

chute. People have used the term Rollerville to describe the flume which had rollers incorporated into the massive lifting device. It's also reported that Rollerville was a different nearby location where the trains circled for reloading with the lumber. *(The local historians need to sort this out.)*

⤳ **MEN 17.2** Call box is located on a sharp curve and overlooks the Garcia River estuary flat. Terrace I dominates the landscape to the north.

⤳ **MEN 18.05** South edge of Garcia River Floodplain, which is nearly .7 of a mile wide at this location.

⤳ **MEN 18.50** Proceed across the Garcia River floodplain to the Garcia River Bridge. During large stream flow periods, the river floods the highway here which is located only 20' above sea level, sometimes closing the road for many hours. There is no other bypass for the flooded area close-by. The stately Snowy Egrets are seen in large flocks on the floodplain here during migration periods.

⤳ **MEN 19.24 Windy Hollow Road** extends south to the casino but is currently not open as a bridge across the Garcia River needs to be reconstructed. The new bridge was planned as part of the construction of the Garcia River Casino, but has not yet been built. This bridge could solve the bypass problem when the river is in flood.

⤳ **MEN 19.34 Mountain View Road** winds its way east 27 miles to Boonville in the Anderson Valley. *(When traveling this winding road it seems like a much longer distance!)* The San Andreas Fault zone is located 1.7 to 2.0 miles to the east of the junction with Highway One. The Point Arena Rancheria is adjacent to the west of the fault trace. Oz Organic Farm straddles the 1906 multi-traces which recorded 15.5' of horizontal movement and 2"–12" of vertical movement down to the west during that event.

⤳ **MEN 19.64 Stoneboro Road** brings you to a must-see destination: the **Manchester Sand Dunes**. You'll drive 1.6 miles across Terrace I at 60'–70' elevation to the dunes parking lot. It's a fairly long, tedious walk over fine-grained sand which makes up the dunes. Reportedly, this is endangered Mountain Beaver habitat, although I have personally never seen one. The dunes are fairly stabilized with dune grasses, however the fierce ocean winds have scooped out blowouts and piled the dunes ever higher. Wildflowers bloom in scattered spots here. There are nice views south to the lighthouse and north along the beach from the top of the closest dunes. The last blowout near the tallest dune has removed all the windblown sand down to the remaining coastal terrace gravel. This gives evidence that the dunes were approximately 20' in height. Manchester State Beach access is located two miles further north and receives many more visitors than this access point; however this area is more pristine as well as more interesting.

⊃   **MEN 20.50 to 20.63** is the hamlet of **Manchester**. Here there's a market, fire station, a business or two, and an elementary school. The trimmed and shaped cypress trees flanking the east side of the highway give Manchester its own distinctive look. Just south of town is a private gravel road, Biaggi Road, which leads to a couple of large ranch homesteads and barn buildings. It has a wide open view across the lower coastal terrace and overlooks an old river flat on the north which is situated just above sea level.

⊃   **MEN 20.62** At the north end of Manchester is a small creek culverted under the highway. This is **Brush Creek** which is located three-quarters of a mile west of the San Andreas Fault at the highway. Brush Creek shows 1.6 miles of offset of its head at the fault. On the east side of the fault, located six-tenths of a mile to the north, the adjacent creek is called **Mill Creek.** This creek shows one mile of offset and then joins Brush Creek on its path to the ocean.

⊃   **MEN 21.40** Turn west on **Kinney Road** to the **KOA Campground**, the **Manchester State Park** coastal access to the **Manchester Dunes**, and the **Manchester Cable Station**, where submarine cables come ashore. Here there's a public restroom at the State Park's parking lot. The park covers 1487 acres onshore and 3782 acres in offshore preserve. The dunes are narrow at this location, and access to the beach is quite easy. The beach extends approximately three miles in length from Alder Creek to the Point Arena Lighthouse.

⊃   MEN 22.48 **Alder Creek Beach Road** is currently closed.

⊃   MEN 22.73 **Alder Creek Bridge** is where the San Andreas Fault leaves the land and goes out to sea. The mouth of Alder Creek is offset approximately one mile to the northwest from its tributaries on the east side of the fault. It appears to be a small stream as it crosses Highway One, but its headwaters extend to the east of the adjacent coastal range. Remember, the majority of these coastal streams are consequent streams which run perpendicular to the rising land and are often only two to five miles in length. The major coastal rivers are few in number and drain large areas of the interior. (See the section on the coastal rivers, page 79.) Alder Creek has a narrow watershed one to two miles wide and extends inland for approximately 11–12 miles, which suggests it's older than the adjacent coastal streams.

⊃   **MEN 23.19** is a private driveway to a residence located on the **North American Plate**. In looking across the valley to the south, it's obvious the terrace on the north side of the fault here is higher than the south terrace. The north terrace, which is Terrace I in appearance, is a broad flat terrace approximately one-half mile in width. The second terrace, Terrace II, is 40' or 50' higher in elevation and is not always preserved as you travel north. Terrace I at bluff edge is 140' elevation rising to 240' elevation east of Highway One.

⊃   **MEN 24.41** **Irish Gulch** is the ravine at the south edge of the community of **Irish Beach.** The short consequent stream with three tributaries cut more than 120' through the coastal terrace into the underlying sandstones of the Franciscan Formation.

⊃   **MEN 25.27** **Vista Point** scenic parking area situated on the ocean bluff. This turn-out is located just north of the entrance on the east to Victoria Gardens, a wonderful historic B&B (bed & breakfast lodging). From the parking area, your view to the south of the nearly vertical bluff edge with an intrepid house at the edge makes one wonder—how strong *are* these rocks? The bluff edge rocks are hard sandstones with nearly vertical dip and so fractured it's difficult to see the bedding planes. The beach here consists of dark grey to black sand, and is not covered with boulders or rock slides that we might expect from looking at the bluff face. The ocean rapidly removes anything that tumbles down this bluff.

The stream at the north edge of Irish Beach is known as **Mallo Pass Creek.** One and one-half miles inland, the stream turns 45º to the south along a northwest-southeast trending fault behind the first coastal ridge. It follows this trend for more than 2.5 miles. A little over one mile north, **Mill Creek** follows the same northwest-southeast trend for the last half mile to the coast.

Probably these creeks are adjusted to old folds and faults in the Franciscan Formation and not adjusted by lateral fault movements.

⊃   **MEN 26.50** **Bridgeport Landing.** An aging barn on the west side of the road is all that's left of this old lumber port. In the Mendocino County historical reports it's recorded a slow-moving landslide crossed and blocked the road the day after the 1906 Earthquake. People observing it at the time thought it was about 100 yards wide, maybe 15' thick, and it progressed several hundred yards. The slide debris was removed off the highway and I can find no evidence of where this event occurred.

⊃ **MEN 26.50 - MEN 30.50** For four miles, look east at the hills and notice there are several steep cuts by streams eroding the hillside terrain. However, most of these waterways do not cross the coastal terrace. There is some evidence of wetland vegetation growing on the terrace, but no stream traces. Instead, the streams sink into porous coastal terrace deposits of sand and gravel. In two areas, successful ponds have been constructed in-line with the suspected underground stream trace. We have noticed this phenomena in several spots along the coast.

⊃ **MEN 30.55 to 31.65** Rapid descent into **Elk Creek**. This is a very strange area. The stream meanders around two or three large, pointed hills or stacks, and Highway One makes a big loop to the east around two of these stacks. Approximately one mile inland, another northwest-southeast regional fault controls the stream pattern, producing a good example of trellis drainage. There may be some lateral movement on this fault. The fault is mapped through the town of **Elk** in two traces, and goes out to sea at **Cuffey's Cove** and just north at **Satori Gulch**.

⊃ **MEN 33.20** **Beacon Light.** An interesting spot for a great view. Restaurant/bar open only Friday and Saturday nights. Situated at approximately 400' elevation on Terrace III and overlooking Terrace I at 200' elevation. Sunset is the ideal time to visit.

⊃ **MEN 33.63** **Greenwood Creek Bridge.** This is a nice, new bridge with an attached pedestrian and bicycle lane screened off from auto traffic. We need more of these types of bridges along the coastal highway.

ELK. This small unincorporated town is currently under a serious reconstruction of its many old buildings. There are restaurants, a Catholic church, gift shops, post office, a museum, B&Bs, an elementary school, and more. At the south end of town is **Greenwood State Beach** which is a must-see. Here cypress trees provide a shady canopy over picnic tables, there's a public restroom, and wide trail(s) lead to the lovely grey-sand beach. It's a great spot for photographing the offshore sea stacks. To the south are two stacks with two sea caves extending through the stacks. There are also many stacks to the north.

⊃ **MEN 35.50** **Cuffey's Cove Cemetery** is located one mile north of Elk. The view is to the south and west over a marvelous group of sea stacks. Incidentally, there are cemetery lots still available, so if you're interested in your final resting place having a splendid view, this might be the place!

We could continue further northward on our journey, but I must stop someplace so I chose this great spot with its memorable vista (after all, a cemetery seems like a logical place to "end"!). If you are continuing north, there are many miles ahead with great views, sea stacks, coastal rivers, and more. The historic and charming towns of **Mendocino** and **Fort Bragg** are not far ahead.

Our journey's end...Cuffey's Cove near the small community of Elk in Mendocino County.

Thanks for your interest in the story of our dynamic coast...

Happy Exploring...!

〰〰

Selected REFERENCES

Alt, David and Hyndman, Donald W., 1975. *Roadside Geology of Northern California.* Mountain Press Publishing Company, Missoula, Montana.

Bailey, Edgar H., Editor. United States Geological Survey,1966. *Geology of Northern California.* Bulletin 190, California Division of Mines & Geology, San Francisco.

Bailey, Edgar H., Irwin, William P., & Jones, David L., 1964. *Franciscan and Related Rocks, and their Significance in the Geology of Western California.* Bulletin 183, California Division of Mines & Geology, Sacramento, California.

Bradley, Raymond S., 2011. *Global Warming and Political Intimidation: How Politicians Cracked Down on Scientists as the Earth Heated Up.* University of Massachusetts Press, Amherst & Boston.

Collier, Michael, 1999. *A Land in Motion, California's San Andreas Fault.* University of California Press, Berkeley, California.

Dott, Robert H. Jr. & Batten, Roger L., 1981. *Evolution of the Earth.* McGraw Hill Book Company, New York.

Elder, William P. Editor, 1998. *Geology & Tectonics of the Gualala Block, Northern California,* Society For Sedimentary Geology, Publication 84.

Erickson, Rolfe, Editor, 1992. *Geology of Northern Marin and Sonoma Counties, California: Three Trips. 1. Pt. Reyes and Vicinity, 2. Geysers Steamfield, 3. Ward Creek-Cazadero Franciscan.* National Association of Geology Teachers, Sonoma State University.

Fagan, Brian, 2013. *The Attacking Ocean, The Past, Present, and Future of Rising Sea Levels.* Bloomsbury Press, New York.

Ferriz, Horacio and Anderson, Robert, Editors, 2001. *Engineering Geology Practice in Northern California,* Bulletin 210, California Geological Survey, Special Publication 12, Association of Engineering Geologists.

Galloway, Alan J., 1977. *Geology of Point Reyes Peninsula, Marin County, California.* California Division of Mines & Geology, Sacramento, California.

Gerhard, Lee C., Harrison, William E. and Hanson, Bernold M., Editors. *Geological Perspectives of Global Climate Change.* AAPG Studies in Geology No. 47. Published by The American Association of Petroleum Geologists in collaboration with the Kansas Geological Survey and the AAPG Division of Environmental Geosciences. Tulsa, Oklahoma.

Griggs, Gary, Patsch, Kiki, and Savoy, Lauret, 2005. *Living with the Changing California Coast,* University of California Press, Berkeley, California.

Hayes, Miles O. and Michel, Jacqueline. 2010. *A Coast to Explore, Coastal Geology and Ecology of Central California.* Pandion Books, Columbia, South Carolina.

Heacox, Kim, 2014. *John Muir and the Ice that Started a Fire.* Lyons Press, Guilford, Connecticut.

Higgins, Charles G., 1952, *Lower Course of the Russian River, California.* University of California Press, Berkeley.

Hill, Mary, 1984. *California Landscape, Origin and Evolution.* University of California Press, Berkeley, California.

Hill, Merton E. III, 2011. *The Concretions of Crystal Cove and Schooner Gulch State Parks and the Anza-Borrego Area.* Saddleback College, Mission Viejo, CA (online publication January 2015).

Howard, Arthur D., 1979. *Geologic History of Middle California.* University of California Press, Berkeley, California.

Huffman, M.E., California Division of Mines and Geology, 1972. *Preliminary Report 16, Geology for Planning on the Sonoma Coast between the Russian and Gualala Rivers.*

Huffman, Michael E., 1973. *Preliminary Report 20. Geology for Planning on the Sonoma Coast Between the Russian River and Estero Americano.* California Division of Mines & Geology and Sonoma County Planning Department.

Huffman, M.E. et al, California Division of Mines & Geology, 1980, *Special Report 120, Geology for Planning in Sonoma County, California.*

Iacopi, Robert, 1964. *Earthquake Country.* Lane Book Company, Menlo Park, California.

Kennedy, Michael P., Greene, H. Gary & Clarke, Samuel H., 1987. *Geology of the California Continental Margin: Explanation of the California Continental Margin Geologic Map Series.* Bulletin 207. U.S. Geologic Survey and the California Coastal Commission. Sacramento, California.

Kolbert, Elizabeth, 2014. *The Sixth Extinction.* Henry Holt and Company, New York.

Konigsmark, Ted, 1994. *Geologic Trips Sea Ranch,.* Self Published, Sea Ranch, California.

Lawson, Andrew C., et al, 1908. *Atlas of Maps and Seismographs Accompanying the Report of the State Earthquake Investigation of April 18, 1908.* Pub. No. 87, Washington, D.C. Carnegie Institute of Washington.

Macdougall, Doug, 2006. *Frozen Earth, The Once and Future Story of Ice Ages.* University of California Press, Berkeley, California.

McPhee, John, 1993. *Assembling California.* Noonday Press, New York.

McPhee, John, 1980. *Basin and Range.* Farrar, Straus & Giroux, New York.

Minard, Claud R., 1971. *Quatenary Beaches and Coasts Between the Russian River and Drakes Bay, California.* College of Engineering, University of California, Berkeley.

Moores, Eldridge M., Editor, 1990. *Shaping the Earth: Tectonics of Continents and Oceans.* Readings from Scientific American Magazine, W.H. Freeman and Company, New York.

Morris, Eugene E,. 1970. *Salt Point State Park Project Report.* California State Department of Education, Santa Rosa, California.

Neild, Ted, 2007. *Supercontinent, Ten Billion Years of Life of our Planet.* Harvard University Press, Cambridge, Massachusetts.

Norris, Robert N. and Webb, Robert W., 1990. *Geology of California. Second Edition.* John Wiley & Sons, Inc. New York, New York.

Patterson, Elizabeth, Project Director, 1993. *California's Rivers, A Public Trust Report.* California State Lands Commission, Sacramento, California.

Prentice, Carol Seabury, 1989. *Earthquake Geology of the Northern San Andreas Fault Near Point Arena, California.* Dissertation: California Institute of Technology, Pasadena, California.

Read, J.F., Kerans, C., and Weber, L.J., 1995. *Milankovitch Sea Level Changes, Cycles and Reservoirs on Carbonate Reservoirs on Carbonate Platforms in Greenhouse and Ice-House Worlds.* Society for Sedimentary Geology, SEPM Short Course Notes No. 35, Tulsa, Oklahoma.

Ruddiman, William F., 2005. *Plows, Plagues, and Petroleum, How Humans Took Control Of Climate.* Princeton University Press, Princeton, N.J.

Schlocker, Julius and Bonilla, M.G., 1964. *Engineering Geology of the Proposed Nuclear Power Plant on Bodega Head, Sonoma County, California.* U.S. Department of the Interior, U.S. Geological Survey, Menlo Park, California.

Sloan, Doris, 2006. *Geology of the San Francisco Bay Region.* California Natural History Guides, University of California Press, Berkeley, California.

Streitz, Robert, and Sherburne, Roger. 1980. *Studies of the San Andreas Fault Zone in Northern California.* Special Report 140, CA Division of Mines & Geology, Sacramento, California.

Turner, Keith A., Schuster, Robert L., Editors, 1996, *Landslides, Investigations and Mitigations, Special Report 247,* Transportation Research Board, National Research Council.

U.S. Department of Agriculture, Forest Service and Soil Conservation Service, 1972, *Soil Survey, Sonoma County, California.*

Wallace, Robert E., Editor, 1990. *The San Andreas Fault System, California.* U.S. Geological Survey Professional Paper 1515. U.S. Government Printing Office, Washington, D.C.

Weaver, Charles E., 1944. *Geology of the Cretaceous (Gualala Group) and Tertiary Formations along the Pacific Coast Between Point Arena and Fort Ross, California.* University of Washington Publications, Volume 6, No. 1, pp1-29, Seattle, Washington.

Wentworth, Carl Merrick, Jr., 1966. *The Upper Cretaceous and Lower Tertiary Rocks of the Gualala Area, Northern Coast Ranges, California.* Dissertation, Stanford University, Palo Alto, California.

Wentworth, C.M., Jr.,1967. *The Upper Cretaceous and Lower Tertiary Rocks of the Gualala Area, Northern Coast Ranges, California. And Inferred Right Slip on the San Andreas Fault.* U.S. Geological Survey, Menlo Park, California.

Winchester, Simon, 2005. *A Crack in the Edge of the World, America and the Great California Earthquake of 1906.* Harper Collins Publishers, New York, New York.

Wright, Terry, 2006. *The Geology of Bodega Head: The Salinian Terrane West of the San Andreas Fault.* Online: TerryWrightGeology.com (April 2, 2015).

INDEX

Alder Creek............6, 8, 27, 33, 42, 57, 59, 65 81, 102, 130

Anchor Bay................18, 23-25, 47, 119, 123

Archaean Eon............................14, 21

Arena Cove.................26, 48-49, 74, 104, 127

Black Point..........22, 24, 44, 46, 90, 98-99 115-117

Black Point Anticline................24-25, 46, 118

Black Point Spilite.......18, 22-24, 44, 46–48 63, 90, 118

Blowholes...................................71

Bodega Bay......6, 22, 32, 35-37, 47, 49, 63, 97

Bowling Ball beach....................26, 71, 73, 125

Cambrian period...........................15

Cape Mendocino...................9, 32, 90-91, 101

Carboniferous Period..............14-15

Cenozoic Era............................16

Chlorite facies...............................23

Coastal Ranges 18, 23, 27, 35, 47, 54, 56 65, 84

Coastal Terraces.........6, 28, 44, 56, 58, 75, 115

Coastal Terrace Deposits........43-44, 63, 70-71 117, 120, 132

Conglomerate 6, 21, 23-24, 46-48, 119

Cretaceous Period..................12-13, 16, 23, 79

Cro-Magnon homo sapiens..........................17

Deep Time........................8-9, 13, 69

Deposition......1, 3, 7, 20, 23-25, 27, 40, 44, 47 59-60, 62, 74-75

Devil's Punchbowl..................74-75

Devonian Period............................15

Elk................60, 62-63, 97, 102, 132

Eel River................................80, 82, 89

Eocene Epoch...............................16

Farallon Plate...............9, 31-32, 102

Fault breccia...................................46

Fort Ross...6, 8, 25, 27, 33, 36, 38-39, 41, 46 57, 64, 110, 112-113

Franciscan Formation..7, 16, 18, 22-23, 27, 35 84-86, 101, 103, 106, 110-111, 131

Galloway Creek Fault...............25, 73

Galloway Formation 18, 25-26, 111, 124

Garcia River....5, 27-28, 33, 42, 48, 59, 70, 79 81, 91, 127-129

Geologic Time Scale.........................14

Geomorphology..............................5

German Rancho Formation....18, 24-25, 47 113, 120

Geysers Geothermal Field.........................9, 83

Climate change..........4, 10, 54, 56, 61-62, 96 107, 122

Gorda Plate...............................32

Greenschist facies..............................47

Gualala..........7-8, 18, 23-3, 35-36, 43-44, 46-48
    56, 62, 64, 81, 89, 90, 117, 120

Gualala River........3, 5, 25, 27-28, 33, 39-43, 46
    48, 60, 63, 71, 79, 81, 83-86, 91-92, 117, 119

Hadean Eon..14

Hathaway Creek Fault............................81, 127

Hole in the head.....................................104-105

Holocene Epoch...17

Igneous dikes & sills...................................7, 23

Igneous rocks..7, 22, 47

Interglacial periods.......................17, 56, 58, 75

Iversen Basalt...............7, 18, 25-26, 47, 124-125

Jenner Grade...7, 23, 59

Jurassic Period..16, 21

Lake Oliver..39-41

Lake Pillsbury..80

Limestone..7, 69

Manchester...................6, 8, 33, 49, 59, 70, 127
    129-130

Mendocino....3, 5, 9, 18, 31-32, 53, 58-59, 65
    69-70, 74, 79-81, 83, 89-91, 121, 133

Mesozoic Era..15

Metamorphic rocks...........................7, 22, 106

Mid-Atlantic Ridge..9

Mill Creek......................37, 41, 111, 130-131

Miller Creek..41-42

Miocene Epoch...16

Mississippian Epoch......................................15

Monterey Canyon..91

Monterey Formation.............18, 26, 49, 74, 123

Navarro River..80

Neanderthals...17

Neogene Period..16

North American Plate....3, 5, 7, 9, 21, 31-32, 48
    56, 65, 131

Noyo Canyon..91-93

Ohlson Ranch Formation.....18, 23, 27, 42, 60
    85-86, 91-92

Ordovician Period...15

Pacific Plate...3, 5, 7-9, 31-32, 36, 42, 48, 56, 60, 65
    74, 81, 102, 105, 127

Paleocene Epoch..16

Paleogene Period..16

Paleozoic Era...15

Pangaea..................................9, 12, 15-16, 101

Pennsylvanian Epoch.....................................15

Permian Period......................................15, 131

Plantation..38-42

Plate tectonics...........................1, 8, 14, 20-21

Pleistocene Epoch..17

Pleistocene Ice Age..................................53, 60

Pliocene Epoch...16

Point Arena........6, 18, 26, 43, 46, 48-50, 57
    63-64, 71, 74-75, 77, 79, 81, 89-90
    92, 102, 104, 125-127-130

Potter Valley..80

Precambrian..14

Proterozoic Eon.................................14

Quaternary Period............................17

Russian River............5, 27-28, 43, 61, 80-81, 83 108-109

Sag ponds.......31, 33, 36, 38-40, 48, 64, 112 117, 121

Salinian block............................32, 35-36, 105

San Andreas Fault Zone.....5, 22-23, 27, 32-33 36, 48, 60, 81, 84, 103, 116, 129

San Gregorio Fault.......................33, 35-36, 105

Sand dunes................6, 35-36, 59, 70, 103-105 120, 129

Schooner Gulch....7, 25-26, 71-72, 74, 124-125

Sea caves..........6, 26, 50, 71, 74-75, 114, 124 128, 132

Sea of Cortez....................................31-32

Sea Stacks.....6, 44, 61-62, 75, 106-107, 109 115, 124, 132

Sedimentary rocks..........................6, 14, 21, 90

Silurian period................................15

Skooner Gulch Formation.......................26, 71

Submarine canyons............................1, 91

Stewarts Point Member...18, 23-24, 47, 116-118

Tafoni..66-67, 114

Tambora volcanic eruption............................55

Terrace I..........18, 27-28, 37, 44, 57-62, 69, 112 115, 129, 131

Terrace II.................................62-63, 115, 132

Terrace III................................63, 132

Terrace IV................................64

Terrace V...................................44, 64

Terrace VI..................................64

Tertiary Period................................109

The Sea Ranch....3-4, 6, 8, 22, 24-25, 27, 33, 44, 46-48, 57, 59-65, 67, 69-70, 81 83, 86, 91, 99, 104, 115-120

Timber Cove.................41, 63, 68, 71, 110, 113

Timber Gulch......................37, 41, 110-111

Triassic Period.........................15, 101

Triplett Gulch..............................25

Tsunami...............................1, 9-10, 24

Turbidity currents.................................6, 91-92

Younger Dryas Glaciation...............................17

Walk-On Beach........10, 25, 59, 62, 70, 119-120

Wisconsinan Glaciation............................17, 58

Questions? Comments? Or otherwise want to get in touch? **Please visit:**

www.RiverBeachPress.com

Or drop a note:

Thomas E. Cochrane
PO Box 358
The Sea Ranch, CA 95497

If you enjoyed **Shaping the Sonoma-Mendocino Coast** *please leave a review on* **Amazon.com**, *and thanks for any other sharing you might do on the book's behalf…*

Meet the author...

Thomas English Cochrane is a California Professional Geologist (License #6124). He lives at The Sea Ranch, California, and has been prowling the landscape of Sonoma and Mendocino Counties since 1976. He was born and raised in Greene, New York, where he roamed and wandered the area's glacially-formed hills and valleys: the Catskills, Adirondacks, and Appalachian mountains. These early explorations led him to acquire an undergraduate degree in Geology from the State University of New York (SUNY), Binghamton. Thereafter, he continued  with graduate studies in Education at Colgate University (NY), graduate work in Geology at Indiana University, and a study in field geology at Miami University's Geology Field Camp in Wyoming.

Pursuing a teaching career, he initially taught science and mathematics as well as earth science and physics in his hometown at Greene Central School for four years, later becoming chairman of the Science Department and acting Chairman of the Mathematics Department.

In 1964, he was the recipient of a National Science Foundation Grant to study Glaciology for a summer on the Juneau Icefield in southeast Alaska.

Mr. Cochrane later spent many years in the oil and gas business, primarily in Oklahoma and Texas. He was editor of the *Shale Shaker,* a geologic publication of the Oklahoma City Geological Society. In 1988, he moved to The Sea Ranch and began consulting on geologic hazards and local geology along the coast. He became a California Registered Geologist in 1995, a professional designation which was recently reclassified as California Professional Geologist.

www.ingramcontent.com/pod-product-compliance
Lightning Source LLC
Chambersburg PA
CBHW061935290426
44113CB00025B/2925